The 1st International Conference of the Red CYTED ENVABIO100 "Obtaining 100% Natural Biodegradable Films for the Food Industry"

The 1st International Conference of the Red CYTED ENVABIO100 "Obtaining 100% Natural Biodegradable Films for the Food Industry"

Editors

Germán Ayala
Mary Lopretti
Maria-Filomena Barreiro
José Vega-Baudrit
Jairo Perilla
Shirley Duarte

Basel • Beijing • Wuhan • Barcelona • Belgrade • Novi Sad • Cluj • Manchester

Editors

Germán Ayala
Federal University of Santa Catarina
Florianópolis, Brazil

José Vega-Baudrit
LANOTEC-CeNAT-CONARE
San José, Costa Rica

Mary Lopretti
Universidad de la República
Montevideo, Uruguay

Jairo Perilla
Universidad Nacional de Colombia
Bogotá, Colombia

Maria-Filomena Barreiro
CIMO-SusTEC, Instituto Politécnico de Bragança
Bragança, Portugal

Shirley Duarte
Universidad Nacional de Asunción
San Lorenzo, Paraguay

Editorial Office
MDPI
St. Alban-Anlage 66
4052 Basel, Switzerland

This is a reprint of articles from the Proceedings published online in the open access journal *Biology and Life Sciences Forum* (ISSN 2673-9976) (available at: https://www.mdpi.com/2673-9976/28/1).

For citation purposes, cite each article independently as indicated on the article page online and as indicated below:

Lastname, A.A.; Lastname, B.B. Article Title. *Journal Name* **Year**, *Volume Number*, Page Range.

ISBN 978-3-0365-9991-5 (Hbk)
ISBN 978-3-0365-9992-2 (PDF)
doi.org/10.3390/books978-3-0365-9992-2

Cover image courtesy of Shirley Duarte

© 2024 by the authors. Articles in this book are Open Access and distributed under the Creative Commons Attribution (CC BY) license. The book as a whole is distributed by MDPI under the terms and conditions of the Creative Commons Attribution-NonCommercial-NoDerivs (CC BY-NC-ND) license.

Contents

Germán Ayala, Maria-Filomena Barreiro, Shirley Duarte, Mary Lopretti, Jairo E. Perilla and José Vega-Baudrit
Preface to the 1st International Conference of the Red CYTED ENVABIO100 "Obtaining 100% Natural Biodegradable Films for the Food Industry"
Reprinted from: *Biol. Life Sci. Forum* **2023**, *28*, 7, doi:10.3390/blsf2023028007 1

José Escurra, Francisco P. Ferreira, Tomás R. López and Walter J. Sandoval-Espinola
Evaluation of Agro-Industrial Carbon and Energy Sources for *Lactobacillus plantarum* M8 Growth
Reprinted from: *Biol. Life Sci. Forum* **2023**, *28*, 1, doi:10.3390/blsf2023028001 3

Talita Szlapak Franco, Graciela Boltzon de Muniz, María Guadalupe Lomelí-Ramírez, Belkis Sulbarán Rangel, Rosa María Jiménez-Amezcua, Eduardo Mendizábal Mijares, et al.
Nanocellulose and Its Application in the Food Industry
Reprinted from: *Biol. Life Sci. Forum* **2023**, *28*, 2, doi:10.3390/blsf2023028002 11

Nicolás Mauricio Bogdanoff, Camilo J. Orrabalis, Wilson Daniel Caicedo Chacon and Germán Ayala Valencia
Study of the Mechanical Properties of Gels Formulated with Pectin from Orange Peel
Reprinted from: *Biol. Life Sci. Forum* **2023**, *28*, 3, doi:10.3390/blsf2023028003 17

Liliana Ávila-Martín, Diana Katherine Guzmán Silva, Jairo E. Perilla and Cristian Camilo Villa Zabala
Preliminary Modeling Study of a Tape Casting System for Thermoplastic Starch Film Forming
Reprinted from: *Biol. Life Sci. Forum* **2023**, *28*, 4, doi:10.3390/blsf2023028004 23

Erika Paulsen, Sofía Barrios and Patricia Lema
Application of Cellulose-Based Film for Broccoli Packaging
Reprinted from: *Biol. Life Sci. Forum* **2023**, *28*, 5, doi:10.3390/blsf2023028005 31

Maria Jaízia dos Santos Alves, Wilson Daniel Caicedo Chacon, Alcilene Rodrigues Monteiro and Germán Ayala Valencia
Starch Nanoparticles Loaded with the Phenolic Compounds from Green Propolis Extract
Reprinted from: *Biol. Life Sci. Forum* **2023**, *28*, 6, doi:10.3390/blsf2023028006 37

Shirley Duarte, Magna Monteiro, Porfirio Andrés Campuzano, Natalia Giménez and María Cristina Penayo
Microcrystals and Microfibers of Cellulose from *Acrocomia aculeata* (Arecaceae) Characterization
Reprinted from: *Biol. Life Sci. Forum* **2023**, *28*, 8, doi:10.3390/blsf2023028008 43

Gabriela Álvarez Véliz, Jorge Iván Cifuentes, Diego Batista, Mary Lopretti, Yendry Corrales, Melissa Camacho and José Roberto Vega-Baudrit
Mechanical Properties of Pineapple Nanocellulose/Epoxy Resin Composites
Reprinted from: *Biol. Life Sci. Forum* **2023**, *28*, 9, doi:10.3390/blsf2023028009 51

Gabriela Lluberas, Diego Batista-Menezes, Juan Miguel Zuñiga-Umaña, Gabriela Montes de Oca-Vásquez, Nicole Lecot, José Roberto Vega-Baudrit and Mary Lopretti
Biofilms Functionalized Based on Bioactives and Nanoparticles with Fungistatic and Bacteriostatic Properties for Food Packing Uses
Reprinted from: *Biol. Life Sci. Forum* **2023**, *28*, 10, doi:10.3390/blsf2023028010 57

Omayra B. Ferreiro and Magna Monteiro
Food Packaging Film Preparation: From Conventional to Biodegradable and Green Fabrication
Reprinted from: *Biol. Life Sci. Forum* **2023**, *28*, 11, doi:10.3390/blsf2023028011 . . . . **69**

Maria Inês Dias, José Pinela, Tânia C. S. P. Pires, Filipa Mandim, Maria-Filomena Barreiro, Lillian Barros, et al.
Biotransformation of Rice Husk into Phenolic Extracts by Combined Solid Fermentation and Enzymatic Treatment
Reprinted from: *Biol. Life Sci. Forum* **2023**, *28*, 12, doi:10.3390/blsf2023028012 . . . . **77**

Shirley Duarte, Axel Dullak, Francisco P. Ferreira, Marcelo Oddone and Darío Riveros
Lactide Synthesis Using ZnO Aqueous Nanoparticles as Catalysts
Reprinted from: *Biol. Life Sci. Forum* **2023**, *28*, 13, doi:10.3390/blsf2023028013 . . . . **87**

Editorial

Preface to the 1st International Conference of the Red CYTED ENVABIO100 "Obtaining 100% Natural Biodegradable Films for the Food Industry"

Germán Ayala [1], Maria-Filomena Barreiro [2,3], Shirley Duarte [4,*], Mary Lopretti [5], Jairo E. Perilla [6] and José Vega-Baudrit [7]

1. Department of Chemical and Food Engineering, Federal University of Santa Catarina, Florianópolis 88000, Brazil; g.ayala.valencia@ufsc.br
2. Research Center of Montanha (CIMO), Polytechnic Institute of Bragança, 5300-253 Bragança, Portugal; barreiro@ipb.pt
3. Associated Laboratory for Sustainability and Technology in the Montanha Regions (SusTEC), Polytechnic Institute of Bragança, Campus of Santa Apolónia, 5300-253 Bragança, Portugal
4. Faculty of Chemistry, National University of Asunción, San Lorenzo 1055, Paraguay
5. Laboratory of Nuclear Techniques Applied to Biochemistry and Biotechnology, Nuclear Research Center, Faculty of Sciences, Universidad de la República, Montevideo 11400, Uruguay; mlopretti@gmail.com
6. Departamento de Ingeniería Química y Ambiental, Universidad Nacional de Colombia, Bogota 111321, Colombia; jeperillap@unal.edu.co
7. National Nanotechnology Laboratory, National Center for High Technology, Pavas 10109, San José, Costa Rica; jvegab@gmail.com
* Correspondence: sduarte@qui.una.py

1. Statement of Peer Review

In submitting these conference proceedings to *Biology and Life Sciences Forum*, the Volume Editors certify to the publisher that all the papers published in this volume have been subjected to peer review administered by the Volume Editors. The reviews were conducted by expert referees to the professional and scientific standards expected in a proceedings journal

- Type of peer review: single-blind
- Conference submission management system: via email
- Number of submissions sent for review: twelve
- Number of submissions accepted: twelve
- Acceptance rate (number of submissions accepted/number of submissions received): one
- Average number of reviews per paper: two
- Total number of reviewers involved: six
- Any additional information on the review process (detailed criteria, peer review policy, etc.): review reports were sent to the authors. All authors made the corrections requested by the reviewers, and only after this were the articles accepted for publication.

2. Conference Description

The event consisted of face-to-face seminars presented to the academic community and the interested general public by researchers who were members of the CYTED ENVABIO100 Network. ENVABIO100 is the Ibero-American network for obtaining biodegradable films of 100% natural origin for the food industry. It is made up of universities, laboratories, and research centers, as well as industries related to the plastics and food sectors (Figure 1). The Network aims to contribute to solutions that lead to the sustainable development of the manufacturing industries and their transition to the use of renewable resources. Over the course of the seminars, a key aim was to cultivate synergy and teamwork between researchers from the region, academia, and the industrial sector of the country, through round tables and visits to factories, in order to achieve the objectives of the Network.

    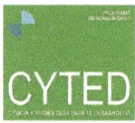

Figure 1. The logos of the institutions that organized or sponsored the conference.

3. Topics

The topics include food sustainability, material technology, packaging for the food industry, and biodegradable polymers.

4. Sponsors

- MADRE SA (https://madre.eco/)
- BIOPLASTIC (https://www.instagram.com/bioplasticpy/)
- CARGILL (https://www.cargill.com.py/)

Conflicts of Interest: The authors declare no conflict of interest.

Disclaimer/Publisher's Note: The statements, opinions and data contained in all publications are solely those of the individual author(s) and contributor(s) and not of MDPI and/or the editor(s). MDPI and/or the editor(s) disclaim responsibility for any injury to people or property resulting from any ideas, methods, instructions or products referred to in the content.

Proceeding Paper

Evaluation of Agro-Industrial Carbon and Energy Sources for *Lactobacillus plantarum* M8 Growth [†]

José Escurra [1], Francisco P. Ferreira [2], Tomás R. López [3] and Walter J. Sandoval-Espinola [4,*]

[1] Plant Biotechnology Laboratory, Department of Biotechnology, Faculty of Exact and Natural Sciences, National University of Asunción, San Lorenzo 111421, Paraguay
[2] Laboratory of Organic Chemistry and Natural Products-LAREV, Department of Biology, Faculty of Exact and Natural Sciences, National University of Asunción, San Lorenzo 111421, Paraguay
[3] Department of Biotechnology, Faculty of Exact and Natural Sciences, National University of Asunción, San Lorenzo 111421, Paraguay
[4] Microbial Biotechnology Laboratory, Department of Biotechnology, Faculty of Exact and Natural Sciences, National University of Asunción, San Lorenzo 111421, Paraguay
* Correspondence: wsandoval@facen.una.py; Tel.: +595-971-298448
[†] Presented at the 1st International Conference of the Red CYTED ENVABIO100 "Obtaining 100% Natural Biodegradable Films for the Food Industry", San Lorenzo, Paraguay, 14–16 November 2022.

Abstract: Lactic acid is a compound used industrially due to its properties. There are two methods for its production: chemical synthesis and microbial fermentation. In microbial fermentation, food industry waste can be used as a substrate, providing a route towards achieving a circular economy. Thus, this study evaluated different substrates for *Lactobacillus plantarum* growth, a lactic acid producer, such as molasses, whey, glucose, and saccharose, either alone or supplemented with additional nutrients. Bacterial growth parameters were assessed using OD_{620} measurement. It was shown that whey supplemented with yeast extract supported the best growth, allowing a $\mu_{max} = 0.63\ h^{-1}$.

Keywords: lactic acid; fermentation; *Lactobacillus*; whey; molasses

1. Introduction

Lactic acid (LA), also known as 2-hydroxypropionic acid (CAS No. 50-21-5), is an organic acid that has been used in food, pharmaceutical, cosmetic, and chemical industries due to its properties as a pH regulator and also as a flavorant, an acidulant, and a preservative. It is also used as an intravascular mineral solution. Currently, the greatest interest in lactic acid is due to it being the precursor of polylactic acid, a biopolymer of great interest today because of its use in bioplastic production [1,2].

Lactic acid can be produced in two ways, with chemical synthesis from petrochemical substrates, and with microbial fermentation, using residues from the food industry as substrates [2,3]. The production of lactic acid using chemical synthesis, in addition to having negative consequences for the environment, has the disadvantage of producing a racemic mixture of (D) and (L) isomers of lactic acid, which makes this method less desirable for industrial use. However, the production of lactic acid through microbial fermentation offers the possibility of obtaining lactic acid with (L) or (D) conformation. The importance of lactic acid conformation (L or D) depends on the industry in which it will be used. In the food and pharmaceutical industry, the L conformation is preferred because it is easily metabolized by humans [3,4]. Other advantages of the production of lactic acid using fermentation are lower costs of substrates, low operating temperatures, and low energy consumption [3].

Lactic acid fermentation is carried out by different microorganisms, including yeasts (*Saccharomyces cerevisiae*, *Candida glycerinogenes*), filamentous fungi (*Aspergillus niger*), and Gram-positive bacteria, including lactic acid bacteria (*Lactobacillus* sp., *Bacillus* sp., and *Enterococcus* sp.) [5–9].

Renewable sources such as starch, lignocellulosic biomass, microalgae, glycerol, and agricultural waste have been proposed as substrates for lactic acid production. The latter substrate has several advantages because it contributes towards achieving a circular economy, that is, the biotransformation of waste into a value-added product [10–14].

To select the best carbon and energy source for lactic acid fermentation, in this study, the growth kinetics of *Lactobacillus plantarum*, using four different substrates obtained from food industry, were evaluated.

2. Materials and Methods

2.1. Activation of the Lactobacillus plantarum Strain on an Erlenmeyer Scale

A stock of the *Lactobacillus plantarum* M8, kindly provided by MSc Yadira Parra from the Department of Biotechnology of the National University of Asunción, was used. In two test tubes, 5 mL of MRS medium was added. Next, 100 µL of its glycerol stock was added to each tube. The tubes were then incubated at 37 °C for 48 h. From the cultures obtained, stocks were prepared in 30% glycerol and were stored at −80 °C until use.

2.2. Analysis of the Growth Kinetics of Lactobacillus plantarum M8 in Different Culture Conditions

In order to determine the best growth condition for *L. plantarum* M8, batch cultures were performed using different carbon sources: food-grade saccharose (7% w/v), glucose (7% w/v), sugarcane molasses (7% v/v), and whey. Molasses and whey were pretreated prior to the growth kinetics test, as described below. Table 1 describes the different culture conditions used in the selected substrates.

Table 1. Substrates evaluated for *Lactobacillus plantarum* M8 growth.

Substrates	Proportion
Glucose	7% (m/v)
Glucose with Yeast Extract	Glucose 7%; 10 g/L yeast extract
Glucose with Meat Peptone	Glucose 7%; 20 g/L meat peptone
Saccharose	7% (m/v)
Saccharose with Yeast Extract	Sucrose 7%; 10 g/L yeast extract
Saccharose with Meat Peptone	7% sucrose; 20 g/L meat peptone
Molasses	7% (v/v)
Molasses with Yeast Extract	Molasses 7%; 10 g/L yeast extract
Molasses with Beef Peptone	Molasses 7%; 20 g/L meat peptone
Whey	Clarified whey
Supplemented Whey	$MgSO_4$ 0.05 g/L; $(NH_4)_2HPO_4$ 2.5 g/L; $MnSO_4$ 0.005 g/L
Whey with Meat Peptone	20 g/L meat peptone
Whey with Yeast Extract	10 g/L yeast extract
Whey Supplemented with Meat Peptone	$MgSO_4$ 0.05 g/L; $(NH_4)_2HPO_4$ 2.5 g/L; $MnSO_4$ 0.005 g/L; 20 g/L meat peptone
Whey Supplemented with Yeast Extract	$MgSO_4$ 0.05 g/L; $(NH_4)_2HPO_4$ 2.5 g/L; $MnSO_4$ 0.005 g/L; 10 g/L yeast extract

2.3. Diluted Molasses Preparation

Under sterile conditions, 25 mL molasses were diluted to 7% (v/v) in sterile distilled water (the molasses was not sterilized) in 50 mL Falcon tubes. The dilutions were centrifuged at 9000 rpm for 5 min, and the supernatant was recovered.

2.4. Whey Pretreatment

Clarification pretreatment using $CaCl_2$ [13]: A 4% (w/v) $CaCl_2$ solution was added to 1000 mL of whey to obtain a concentration of 0.02% (v/v) and autoclaved under standard conditions for 15 min. This mixture was then centrifuged at 9000 rpm for 5 min, and the supernatant was recovered. Small colloids that remained were filtered out via filter paper. The clarified whey was then sterilized and stored at 4 °C until use.

2.5. Activation of the Lactobacillus plantarum M8

From *Lactobacillus plantarum* M8 glycerol stocks, 200 µL was taken and transferred to a 5 mL MRS-Broth medium in triplicate, incubated at 37 °C for 19 h. Subsequently, 10 mL of the previously activated strains were inoculated into 200 mL of fresh MRS-Broth medium, and cultured at 37 °C, at 150 rpm. After 28 h, the optical density was determined, and the calculation was performed to start the experimental cultures with an optical density of 0.1.

2.6. Lactobacillus plantarum M8 Growth Kinetics under Different Carbon Sources

From the solutions prepared (saccharose, glucose, molasses, and whey) and the MRS-Broth medium, which was used as control, a 1 mL aliquot was taken and inoculated with the activated strain of *L. plantarum*. Then, 200 µL of each inoculum was transferred to a 96-well plate, in triplicate. Culture medium without cells were used as blanks. Each culture's growth was monitored via optical density measurement at 620 nm (OD620 nm) for 24 h, via plate reader at 37 °C (Multiskan, Thermo Fisher Scientific, Waltham, MA, USA), with pulse shakes before each reading, which was automatically performed every 30 min, along with each reading. Optical density is an indirect measurement of microbial growth, typically used in fermentation assays. As microbes proliferate, the sample's optical density increases linearly until a certain value, usually 0.9, after which samples must be diluted so that the linearity of optical density continues [15]. In our experiments, linearity was maintained with corrections performed automatically by the plate reader.

3. Results and Discussion

3.1. Growth Curves and Biomass Concentration

The data obtained from the growth kinetics of *L. plantarum* M8 using different substrates were analyzed. Figure 1 shows the growth curve in the different glucose conditions, obtaining higher OD620 nm in glucose with meat peptone, followed by glucose with yeast extract. Glucose without supplementation did not support growth, as expected due to the lack of nutrients. Specifically, this may be due to the lack of nitrogen sources and other micro- and macronutrients.

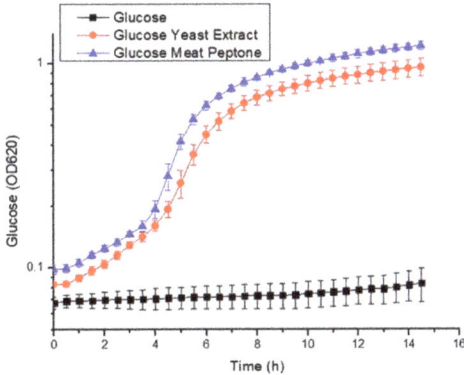

Figure 1. Growth kinetics of *L. plantarum* M8 under different glucose supplementation conditions.

With respect to sucrose, Figure 2 shows that saccharose supplemented with meat peptone and yeast extract were similar. However, the meat peptone's culture started the exponential growth phase approximately 1 h earlier.

Along with glucose and saccharose, molasses, which is an agro-industrial waste, was also evaluated as a substrate. For this evaluation, molasses was diluted to 7% (v/v) prior to supplementation, as described in Materials and Methods. Figure 3 shows *L. plantarum* M8 growth under different molasses supplementation conditions. Molasses without supplementation presented a lower growth than the other two conditions, but a higher growth compared to the growth obtained with glucose and sucrose, both without supplementation. This may be due to the presence of other limiting elements in molasses, including nitrogen sources and other nutrients. On the other hand, molasses with yeast extract, and molasses with meat peptone, produced similar optical densities. This might indicate the presence of the same limiting elements, and the potential for a higher growth under controlled fermentation conditions.

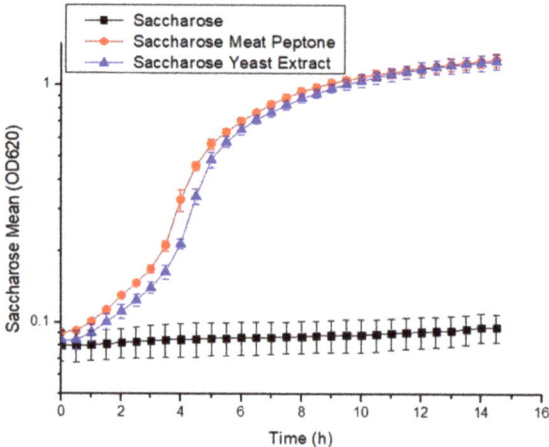

Figure 2. Growth kinetics of *L. plantarum* M8 under different saccharose supplementation conditions.

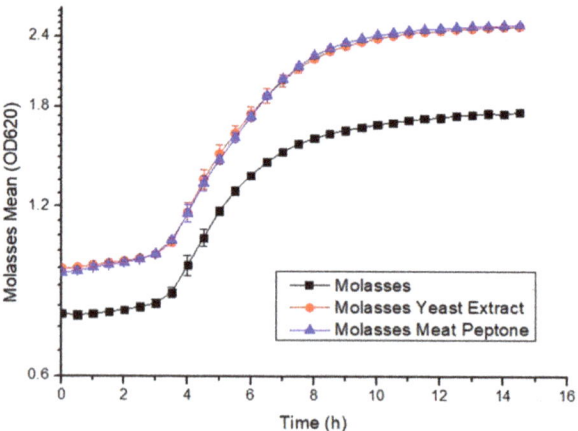

Figure 3. Growth kinetics of *L. plantarum* under different molasses conditions.

Whey, which is another agro-industrial waste, was also evaluated. This substrate was subjected to a clarification process prior to its use, as described in Materials and Methods. Figure 4 shows the growth curves of *L. plantarum* M8 under different whey conditions. The condition containing yeast extract presented the highest final OD620 nm. Neither whey supplemented with salts (red symbols) nor with salts and yeast extract (green symbols) supported microbial growth, as inferred from the obtained flat curves and the formation of precipitates (i.e., high initial OD620).

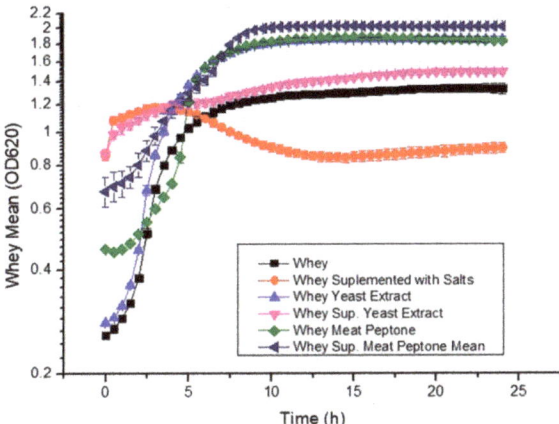

Figure 4. Growth kinetics of *L. plantarum* under different whey conditions.

3.2. Maximum Specific Growth Grates (μ_{Max}) Calculations

The results obtained from the growth kinetics of *L. plantarum* were linearized in order to calculate the maximum specific growth rate (μ_{max}). Table 2 shows the μ_{max} values obtained in the different culture conditions.

Table 2. Maximum growth rate of *L. plantarum*.

Condition	μ_{max} (h^{-1})
MRS	0.67
Molasses	0.2246
Molasses with Beef Peptone	0.2326
Molasses with Yeast Extract	0.26
Saccharose	0
Sucrose with Meat Peptone	0.2519
Sucrose with Yeast Extract	0.7258
Glucose	0
Glucose with Meat Peptone	0.7714
Glucose with Yeast Extract	0.216
Whey	0.59
Whey Supplemented with Salts	0.027
Whey with Yeast Extract	0.63
Whey with Meat Peptone	0.167
Supplemented Whey + Yeast Extract	0.046
Supplemented Whey + Meat Peptone	0.193
MRS Broth	0.65

The μ_{max} obtained in the MRS medium was 0.67 h^{-1}, and this was our study control. The condition that presented the best growth was whey with yeast extract, showing a μ_{max} = 0.63 h^{-1}, followed by whey without supplementation (μ_{max} = 0.59 h^{-1}). With molasses, growth rates were slower compared to the other conditions, although they supported a high final OD6020. Between glucose and sucrose conditions, glucose supplemented with meat peptone had the highest growth rate at μ_{max} = 0.7714 h^{-1}.

Overall, these results suggest that the best growth conditions for *L. plantarum* M8 are whey with yeast extract and unsupplemented whey. These results coincide with those reported by other authors, where both unsupplemented and supplemented whey also present higher lactic acid concentration, volumetric productivity, and yield during fermentation [14,16,17]. In this case, growth is a good proxy for lactic acid production, because this organic acid drives ATP generation and thus growth in lactic acid bacteria.

4. Conclusions

Lactic acid generation using microbial cultures has a potential for supplying this commodity for polylactic acid production or bioplastic production. Considering that the carbon and energy sources compatible with lactic acid bacteria (LAB) growth can be agricultural byproducts, both the bioplastic itself and the process for generating its molecular scaffold are compatible with sustainable industrial practices, including achieving a circular economy. Understanding LAB growth kinetics under these conditions will allow us to achieve better process development and the subsequent optimization of lactic acid production, in terms of yield, concentration, and productivity.

Author Contributions: Conceptualization, J.E., T.R.L. and W.J.S.-E.; methodology, J.E., F.P.F. and W.J.S.-E.; software W.J.S.-E. and J.E.; validation, J.E., F.P.F. and W.J.S.-E.; formal analysis, J.E. and W.J.S.-E.; resources, W.J.S.-E.; data curation, W.J.S.-E. and J.E.; writing—J.E. and W.J.S.-E.; writing—review and editing: J.E. and W.J.S.-E.; visualization, J.E. and W.J.S.-E.; supervision, W.J.S.-E.; project administration, W.J.S.-E. and T.R.L.; funding acquisition: T.R.L. and W.J.S.-E. All authors have read and agreed to the published version of the manuscript.

Funding: This research was partially funded by Red Cyted ENVABIO100 121RT0108 and by the Faculty of Natural and Exact Sciences of the National University of Asuncion.

Institutional Review Board Statement: Not applicable.

Informed Consent Statement: Not applicable.

Data Availability Statement: Not applicable.

Conflicts of Interest: The authors declare no conflict of interest.

References

1. Komesu, A.; de Oliveira, J.A.R.; da Silva Martins, L.H.; Maciel, M.R.W.; Maciel Filho, R. Lactic acid production to purification: A review. *BioResources* **2017**, *12*, 4364–4383. [CrossRef]
2. Cubas-Cano, E.; González-Fernández, C.; Ballesteros, M.; Tomás-Pejó, E. Biotechnological advances in lactic acid production by lactic acid bacteria: Lignocellulose as novel substrate. *Biofuels Bioprod. Biorefin.* **2018**, *12*, 290–303. [CrossRef]
3. de Albuquerque, T.L.; Junior, J.E.M.; de Queiroz, L.P.; Ricardo, A.D.S.; Rocha, M.V.P. Polylactic acid production from biotechnological routes: A review. *Int. J. Biol. Macromol.* **2021**, *186*, 933–951. [CrossRef] [PubMed]
4. Wee, Y.J.; Kim, J.N.; Ryu, H.W. Biotechnological production of lactic acid and its recent applications. *Food Technol. Biotechnol.* **2006**, *44*, 163–172.
5. Wang, Y.; Chen, C.; Cai, D.; Wang, Z.; Qin, P.; Tan, T. The optimization of L-lactic acid production from sweet sorghum juice by mixed fermentation of *Bacillus coagulans* and *Lactobacillus rhamnosus* under unsterile conditions. *Bioresour. Technol.* **2016**, *218*, 1098–1105. [CrossRef] [PubMed]
6. Ge, X.Y.; Qian, H.; Zhang, W.G. Improvement of L-lactic acid production from Jerusalem artichoke tubers by mixed culture of *Aspergillus niger* and *Lactobacillus* sp. *Bioresour. Technol.* **2009**, *100*, 1872–1874. [CrossRef] [PubMed]
7. Ouyang, J.; Ma, R.; Zheng, Z.; Cai, C.; Zhang, M.; Jiang, T. Open fermentative production of L-lactic acid by *Bacillus* sp. strain NL01 using lignocellulosic hydrolyzates as low-cost raw material. *Bioresour. Technol.* **2013**, *135*, 475–480. [CrossRef] [PubMed]
8. Baek, S.H.; Kwon, E.Y.; Bae, S.J.; Cho, B.R.; Kim, S.Y.; Hahn, J.S. Improvement of d-lactic acid production in *Saccharomyces cerevisiae* under acidic conditions by evolutionary and rational metabolic engineering. *Biotechnol. J.* **2017**, *12*, 1700015. [CrossRef] [PubMed]
9. Sun, Y.; Xu, Z.; Zheng, Y.; Zhou, J.; Xiu, Z. Efficient production of lactic acid from sugarcane molasses by a newly microbial consortium CEE-DL15. *Process Biochem.* **2019**, *81*, 132–138. [CrossRef]
10. Abdel-Rahman, M.A.; Hassan, S.E.D.; El-Din, M.N.; Azab, M.S.; El-Belely, E.F.; Alrefaey, H.M.A.; Elsakhawy, T. One-factor-at-a-time and response surface statistical designs for improved lactic acid production from beet molasses by *Enterococcus hirae* ds10. *SN Appl. Sci.* **2020**, *2*, 573. [CrossRef]
11. Liu, P.; Zheng, Z.J.; Xu, Q.Q.; Qian, Z.J.; Liu, J.H.; Ouyang, J. Valorization of dairy waste for enhanced D-lactic acid production at low cost. *Process Biochem.* **2018**, *71*, 18–22. [CrossRef]
12. Juturu, V.; Wu, J.C. Microbial production of lactic acid: The latest development. *Crit. Rev. Biotechnol.* **2016**, *36*, 967–977. [CrossRef] [PubMed]
13. Fernando Escobar, L.; Andres Rojas, C.; Giraldo, G.; Antonio, G.; Padilla Sanabria, L. Evaluation of lactobacillus casei growth and production of lactic acid using as substrate the whey of bovine milk. *Res. J.-Univ. Quindio* **2010**, *20*, 42–49.
14. Lech, M. Optimization of protein-free waste whey supplementation used for the industrial microbiological production of lactic acid. *Biochem. Eng. J.* **2020**, *157*, 107531. [CrossRef]

15. Sandoval-Espinola, W.J.; Chinn, M.; Bruno-Barcena, J.M. Inoculum optimization of *Clostridium beijerinckii* for reproducible growth. *FEMS Microbiol. Lett.* **2015**, *362*, fnv164. [CrossRef] [PubMed]
16. Bernardo, M.P.; Coelho, L.F.; Sass, D.C.; Contiero, J. L-(+)-Lactic acid production by *Lactobacillus rhamnosus* B103 from dairy industry waste. *Braz. J. Microbiol.* **2016**, *47*, 640–646. [CrossRef] [PubMed]
17. Luongo, V.; Policastro, G.; Ghimire, A.; Pirozzi, F.; Fabbricino, M. Repeated-batch fermentation of cheese whey for semi-continuous lactic acid production using mixed cultures at uncontrolled pH. *Sustainability* **2019**, *11*, 3330. [CrossRef]

Disclaimer/Publisher's Note: The statements, opinions and data contained in all publications are solely those of the individual author(s) and contributor(s) and not of MDPI and/or the editor(s). MDPI and/or the editor(s) disclaim responsibility for any injury to people or property resulting from any ideas, methods, instructions or products referred to in the content.

Proceeding Paper

Nanocellulose and Its Application in the Food Industry [†]

Talita Szlapak Franco [1], Graciela Boltzon de Muniz [1], María Guadalupe Lomelí-Ramírez [2], Belkis Sulbarán Rangel [3], Rosa María Jiménez-Amezcua [4], Eduardo Mendizábal Mijares [5], Salvador García-Enríquez [2,*] and Maite Rentería-Urquiza [5,*]

[1] Department of Forest Engineering, Federal University of Parana, Curitiba 80210-170, Brazil; talitaszlapak@gmail.com (T.S.F.); gbmuniz@ufpr.br (G.B.d.M.)
[2] Department of Wood Cellulose and Paper, University of Guadalajara, Guadalajara 44430, Mexico; maria.lramirez@academicos.udg.mx
[3] Department of Water and Energy Studies, University of Guadalajara, Guadalajara 44430, Mexico; belkis.sulbaran@academicos.udg.mx
[4] Department of Chemical Engineering, University of Guadalajara, Guadalajara 44430, Mexico; rosa.jamezcua@academicos.udg.mx
[5] Department of Chemistry, University of Guadalajara, Guadalajara 44430, Mexico; eduardo.mmijares@academicos.udg.mx
* Correspondence: salvador.genriquez@academicos.udg.mx (S.G.-E.); maite.rurquiza@academicos.udg.mx (M.R.-U.)
[†] Presented at the 1st International Conference of the Red CYTED ENVABIO100 "Obtaining 100% Natural Biodegradable Films for the Food Industry", San Lorenzo, Paraguay, 14–16 November 2022.

Abstract: This work presents a review related to the obtainment of cellulose from different structures in agro-industrial residues, both for application in the food industry and for the reinforcement of other materials. Cellulose nanofibers are produced by the heart of palm (*Bactris gasipaes*) industry in Brazil and are used as a stabilizer in avocado oil emulsions; conversely, cellulose nanocrystals are produced in waste from the tequila industry (*Agave tequilana* Weber var. Azul) in Jalisco, Mexico, and are used for reinforcement applications.

Keywords: cellulose nanofibrils; cellulose nanocrystals; agro-industrial residues; *Bactris gasipaes*; *Agave tequilana* Weber

1. Introduction

Cellulose is considered a natural polymer of great abundance, since it is possible to obtain it from very diverse sources such as animals; microorganisms; non-timber fibers such as resins, gums, and waxes; and others (fungi, seeds, leaves, nopal, stems, fruits, etc.) [1]. Cellulose can be obtained from plants in their virgin state and from the waste they themselves generate. In this way, a material that would otherwise be discarded, and that could cause environmental problems, is revalued [2].

The name nanocellulose refers to the nanometric-scale dimensions of this natural polymer. There are three types of nanocellulose, categorized depending on their production and extraction: crystal-shaped nanocellulose (NCC), nanocellulose fibers (NFC), and bacterial nanocellulose (NCB) [3].

The generation of new materials based on nanocellulose has become an increasingly attractive area of development, because these nanomaterials have the characteristics of sustainability, biodegradability, non-toxicity, and economic production [4].

Applications of nanomaterials include important industries such as paper, food, electronics, pharmaceutical, biomedical engineering, construction, packaging, etc. [5,6].

Despite the fact that all of these industries generate waste, the food industry is one of the sectors that generates the most environmental impact, due to its processes and the different products they generate [7]. Therefore, finding an application for waste generated in industries like this is a challenge, but a necessity for the achievement of a sustainable society.

2. Lignocellulosic Waste

Due to the large-scale production of tequila in the state of Jalisco in Mexico, agave bagasse is an abundant source of lignocellulosic biomass [8–10]. Bagasse is a solid by-product of a fibrous nature, obtained after the grinding of the agave pineapple and the extraction of fermentable sugars in the manufacture of tequila [8]. It is estimated that the fibrous biomass resulting from the grinding of agave pineapple is equivalent to 40% of the total wet weight [11]. Agave bagasse, due to its high availability, has traditionally always presented serious problems for the industry, because its final disposal comes at high management costs [12]. This has led to the problem of environmental contamination, because most of this waste ends up as waste in clandestine dumps due to a lack of environmental regulation. This causes negative effects on the fertility of farmland [13], leachate contamination, and phytosanitary risks due to the inadequate incorporation of this material into soil [12,14]. However, based on its chemical composition, it is known that agave bagasse contains 44.5% cellulose, 25.3% hemicellulose, and; 20.1% lignin [9,15], so researchers have tried to diversify its applications in different areas, such as the production of biopolymers, composting, animal feed, and the generation of biofuels [8,11] and reinforcement materials. It has recently been studied in relation to the production of nanofibers and cellulose nanocrystals [9,10,15,16].

In the case of the of peach palm heart production in Brazil, where the economy is most reliant on agriculture, the exploitation of different lignocellulosic residues and wastes for nanocellulose production offers a great chance to increase the income of small-scale companies and farmers, and to achieve the sustainable develop- ment of agriculture. The extraction of peach palm from *Bactris gasipaes* (aka pupunha) palm trees to produce peach palm heart (palmito) produces high amounts of residues, since just 10% of the tree is used for food production, and the other parts generally are used in farming activities or energy production. Brazil is the major producer and consumer of palmito, and the waste from their plantation in 2017 represented approximately 5×10^5 tons of cellulose that was released into the field or incinerated, but could have been used for more justifiable and profitable purposes [17,18]. Figure 1 shows the scheme of a method for obtaining nanofibers from palm residues and nanocrystals from agave bagasse.

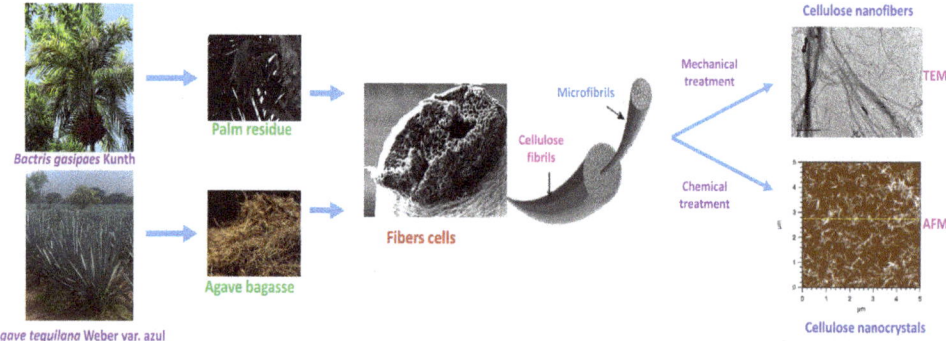

Figure 1. Obtaining cellulose nanofibers and nanocrystals from palm and agave residues.

3. Characterization of Cellulose Nanostructures

Lignocellulosic materials are characterized by the presence of cell walls, mainly made up of a series of coaxial layers of cellulose microfibrils (skeleton) dispersed in an amorphous matrix of hemicelluloses and lignin, which together represent 80–90% of the total weight [19]. Among the structural components of the cell wall are lignin and polysaccharides, the most abundant being cellulose and hemicelluloses.

Nanomaterials or nanometric materials have attracted scientific interest in recent years because they have better properties, whether electrical, mechanical, thermal, etc., than materials with the same composition but a macrometric size. By definition, these must have at least one of their dimensions in the range of 0.1 to 100 nm, although some authors consider them up to 600 nm [20].

The extraction and production of nanocellulose from various sources has attracted increasing interest due to this material's abundance, strength, rigidity, low weight, and biodegradability [21]. Different terms are used in the literature to designate these cellulose nanoparticles in the form of crystalline rods. They are mainly referred to as whiskers, nanowhiskers (NWC), nanofibers (NFC), cellulose nanocrystals (NCC), monocrystals, and microcrystals [22]. However, the dimensions of cellulosic nanoparticles depend on several factors, including the source of cellulose and the exact preparation conditions [23].

Through different mechanical, chemical, enzymatic or biological processes, it is possible to obtain nanofibers (NFC) and cellulose nanocrystals (NCC), which are the most basic structural forms of this polysaccharide. They have crystalline domains, which have excellent mechanical properties, and an elastic modulus of the order of 150 GPa, which is greater than the elastic modulus of glass fibers (85 GPa) and that of aramid (65 GPa) [24]. Therefore, these nanomaterials can provide considerable improvements to the mechanical properties of the matrices to which they are added. NFCs an elongated cylindrical shape with a high aspect ratio; they are very long in relation to their diameter. NCCs take the shape of an elongated grain of rice. These nanomaterials have high potential for use in multiple ways, particularly as reinforcing materials for the development of nanocomposites, due to the fact that they have a large specific surface area, that is, the area per unit mass wherein they can interact directly with the matrix [20]. Many studies have been performed for the isolation and characterization of NFC and NCC from various sources. This is why the elaboration of nanocomposites with replacement capacity, to act as regenerative agents and as structural supports for various materials, is a viable option [25]. In this sense, it is also hoped that these materials will be environmentally friendly, that they will allow production on an industrial scale [26], and that they will be of biological origin. All of the above factors will give them a comparative advantage over conventional materials, such as ceramics, metals, and polymers [27].

4. Nanocellulose in Food

Nanomaterials are used in the food industry to improve the quality of food products. They can prevent microbial degradation of packaged foods, improve their color, flavor or texture, and increase the bioavailability of vitamins and minerals [28,29].

Nanomaterials used in food can be classified into three different groups [30]:

Organic nanomaterials include lipids, proteins, and polysaccharides, which are used to encapsulate vitamins, antioxidants, dyes, flavorings and preservatives, and form micelles, liposomes, and nanospheres, etc. They allow for higher intake, absorption, bioavailability, and stability in the body.

Organic/inorganic combined nanomaterials are also called surface functionalized nanomaterials, generally added to a matrix for their specific functionality (antimicrobials, antioxidants, and permeability and rigidity regulators).

Inorganic nanomaterials are metals and metal oxides of Ag, Fe, Se, TiO_2, used as additives, food supplements, or in packaging [31].

Avocado is one of the most abundant fruits in Mexico. In fact, Mexico is the biggest exporter of avocado in the world. There are many studies that show that consumption of avocado provides optimal fats human needs [32]. Avocado oil, due to its fatty acid composition, meets nutritional recommendations that focus on reducing the amount of saturated fat in the diet [32–34]. Diets rich in avocado oil have been shown to be effective in reducing total cholesterol, LDL (low-density lipoprotein) cholesterol, and plasma triglycerides, as is the case with diets containing corn, soybean, or sunflower oil [32].

Avocado oil-based emulsions are widely used as a food supplement or dressing because they are healthier than alternatives. More and more methods of extending the expiration and stability of these emulsions are sought [35–37].

The use of cellulose nanofibers (NFC) to improve some properties in food can be carried out in three different ways: via food additives (food supplements), in packaging, or in emulsifiers [38]. In this case, their effects can be compared with those produced by other emulsifiers commonly used in this sector, such as sorbitan monostearate or Span 60 [39]. This behavior was reported by Talita et al. in a study of the stabilization of avocado oil emulsions with cellulose nanofibers obtained from the palpito (*Bactris gasipaes*) industry waste in Brazil [18]. These nanometric structures allowed the elimination of the phenomenon of coalescence in emulsions, and the formation of cream on the surface of them [18]. In addition, it was found that the stability of the emulsions created increased, as was the same as that of emulsions that were refrigerated. The particle size of the micelles formed decreased as the percentage of considered cellulose nanofibers increased, which allowed the researchers to establish the influence of these nanostructures on the final stability of the system.

5. Nanocellulose as Reinforcement

The use of cellulose as a reinforcement material for synthetic polymers and recycled plastic materials has generated great interest. There are many studies related to the analysis of the mechanical properties of biocomposites obtained via mixing both materials [40,41]. It has been possible to prove that the addition of nanocellulose to synthetic polymers significantly favors some of the properties of these polymers, such as tensile strength and thermal conductivity; however, this depends on the initial nature of the cellulose [42–44].

In studies about the reinforcement of polylactic acid (PLA) with nanocellulose, it has been verified that this combination of materials has the potential to be competitive, since the properties in general were similar to those of petroleum-derived polymers. Both the polymeric matrix and the reinforcement additives made it possible to achieve biodegradable materials, which were approved for contact with food [45]. Pech et al. reported that in using cellulose nanocrystals from tequila industry waste (residues of *Agave tequilana* Weber var. Blue), it is possible to observe this behavior [46].

6. Conclusions

In this paper, we present two direct applications of nanocellulose in the form of nanofibers and nanocrystals, both in the food industry and as a reinforcement for other materials.

This is a small sample of what it is possible to do with this biopolymer in pursuit of global sustainability. The use of the lignocellulosic waste that each region of the planet produces on a regular basis can be made into a source of income for the most disadvantaged populations. It is possible to improve food properties with the gradual incorporation of cellulose nanostructures, although we must be cautious about the side effects they can cause.

Author Contributions: Conceptualization, R.M.J.-A. and M.R.-U.; methodology, S.G.-E. and T.S.F.; formal analysis, E.M.M. and M.G.L.-R.; investigation, M.R.-U., G.B.d.M. and B.S.R.; resources, T.S.F., S.G.-E. and M.R.-U.; writing—original draft preparation, M.R.-U. and S.G.-E.; writing—review and editing, S.G.-E. and M.R.-U.; visualization, S.G.-E.; supervision, M.R.-U.; project administration, R.M.J.-A.; funding acquisition, R.M.J.-A. All authors have read and agreed to the published version of the manuscript.

Funding: This research was funded by Red Cyted ENVABIO100 121RT0108.

Institutional Review Board Statement: Not applicable.

Informed Consent Statement: Not applicable.

Data Availability Statement: Not applicable.

Conflicts of Interest: The authors declare no conflict of interest.

References

1. Dufresne, A. Nanocellulose: A new ageless bionanomaterial. *Mater. Today* **2013**, *16*, 220–227. [CrossRef]
2. Ilyas, R.A.; Sapuan, S.M.; Ibrahim, R.; Atikah, M.S.N.; Atiqah, A.; Ansari, M.N.M.; Norrrahim, M.N.F. Production, processes and modification of nanocrystalline cellulose from agro-waste: A review. In *Nanocrystalline Materials*; IntechOpen: London, UK, 2019; pp. 3–32.
3. Omran, A.A.B.; Mohammed, A.A.; Sapuan, S.M.; Ilyas, R.A.; Asyraf, M.R.M.; Rahimian Koloor, S.S.; Petrů, M. Micro-and nanocellulose in polymer composite materials: A review. *Polymers* **2021**, *13*, 231. [CrossRef] [PubMed]
4. Chaker, A.; Mutjé, P.; Vilar, M.R.; Boufi, S. Agriculture crop residues as a source for the production of nanofibrillated cellulose with low energy demand. *Cellulose* **2014**, *21*, 4247–4259. [CrossRef]
5. Abitbol, T.; Rivkin, A.; Cao, Y.; Nevo, Y.; Abraham, E.; Ben-Shalom, T.; Lapidot, S.; Shoseyov, O. Nanocellulose, a tiny fiber with huge applications. *Curr. Opin. Biotechnol.* **2016**, *39*, 76–88. [CrossRef]
6. Curvello, R.; Raghuwanshi, V.S.; Garnier, G. Engineering nanocellulose hydrogels for biomedical applications. *Adv. Colloid Interface Sci.* **2019**, *267*, 47–61. [CrossRef] [PubMed]
7. Macedo, M.E. Microfibrillated Cellulose and High-Value Chemicals from Orange Peel Residues. Ph.D. Thesis, University of York, York, UK, 2018.
8. Alemán-Nava, G.S.; Gatti, I.A.; Parra-Saldivar, R.; Dallemand, J.F.; Rittmann, B.E.; Iqbal, H.M. Biotechnological revalorization of Tequila waste and by-product streams for cleaner production–A review from bio-refinery perspective. *J. Clean. Prod.* **2018**, *172*, 3713–3720. [CrossRef]
9. Hernández, J.A.; Romero, V.H.; Escalante, A.; Toríz, G.; Rojas, O.J.; Sulbarán, B.C. *Agave tequilana* bagasse as source of cellulose nanocrystals via organosolv treatment. *Bioresources* **2018**, *13*, 3603–3614. [CrossRef]
10. Robles-García, M.Á.; Del-Toro-Sánchez, C.L.; Márquez-Ríos, E.; Barrera-Rodríguez, A.; Aguilar, J.; Aguilar, J.; Reynoso-Marín, F.J.; Ceja, I.; Dorame-Miranda, R.; Rodríguez-Félix, F. Nanofibers of cellulose bagasse from *Agave tequilana* Weber var. azul by electrospinning: Preparation and characterization. *Carbohydr. Polym.* **2018**, *192*, 69–74. [CrossRef] [PubMed]
11. Iñiguez-Covarrubias, G.; Lange, S.E.; Rowell, R.M. Utilization of byproducts from the tequila industry: Part 1: Agave bagasse as a raw material for animal feeding and fiberboard production. *Bioresour. Technol.* **2001**, *77*, 25–32. [CrossRef]
12. Zamora Natera, F.; Ruiz López, M.A.; García López, P.M.; Rodríguez Macías, R.; Iñiguez Covarrubias, G.; Salcedo Pérez, E.; Alcantar González, E.G. Caracterización física y química de sustratos agrícolas a partir de bagazo de agave tequilero. *Interciencia* **2010**, *35*, 515–520. Available online: https://www.interciencia.net/wp-content/uploads/2018/01/515-c-RODR%C3%8DGUEZ-MAC%C3%8DAS-7.pdf (accessed on 12 May 2023).
13. Gobeille, A.; Yavitt, J.; Stalcup, P.; Valenzuela, A. Effects of soil management practices on soil fertility measurements on *Agave tequilana* plantations in Western Central Mexico. *Soil Tillage Res.* **2006**, *87*, 80–88. [CrossRef]
14. Zurita, F.; Tejeda, A.; Montoya, A.; Carrillo, I.; Sulbarán-Rangel, B.; Carreón-Álvarez, A. Generation of tequila vinasses, characterization, current disposal practices and study cases of disposal methods. *Water* **2022**, *14*, 1395. [CrossRef]
15. Palacios Hinestroza, H.; Hernández Diaz, J.A.; Esquivel Alfaro, M.; Toriz, G.; Rojas, O.J.; Sulbarán-Rangel, B.C. Isolation and Characterization of Nanofibrillar Cellulose from *Agave tequilana* Weber Bagasse. *Adv. Mater. Sci. Eng.* **2019**, *2019*, 1342547. [CrossRef]
16. Lomelí-Ramírez, M.G.; Valdez-Fausto, E.M.; Rentería-Urquiza, M.; Jiménez-Amezcua, R.M.; Anzaldo Hernández, J.; Torres-Rendon, J.G.; García Enriquez, S. Study of green nanocomposites based on corn starch and cellulose nanofibrils from *Agave tequilana* Weber. *Carbohydr. Polym.* **2018**, *201*, 9–19. [CrossRef] [PubMed]
17. Franco, T.S.; Potulski, D.C.; Viana, L.C.; Forville, E.; de Andrade, A.S.; de Muniz, G.I.B. Nanocellulose obtained from residues of peach palm extraction (*Bactris gasipaes*). *Carbohydr. Polym.* **2019**, *218*, 8–19. [CrossRef]
18. Franco, T.S.; Rodríguez, D.C.M.; Soto, M.F.J.; Amezcua, R.M.J.; Urquíza, M.R.; Mijares, E.M.; de Muniz, G.I.B. Production and technological characteristics of avocado oil emulsions stabilized with cellulose nanofibrils isolated from agroindustrial residues. *Colloids Surf. A Physicochem. Eng. Aspects* **2020**, *586*, 124263. [CrossRef]
19. Huang, J.; Ma, X.; Yang, G.; Alain, D. Introduction to nanocellulose. In *Nanocellulose: From Fundamentals to Advanced Materials*; Wiley Online Library: Hoboken, NJ, USA, 2019; pp. 1–20. [CrossRef]
20. Riva, G.H.; Silva, J.A.; Navarro, F.; López-Dellamary, F.; Robledo, J.R. Síntesis de nanocompuestos de celulosa para aplicaciones biomédicas en base a sus propiedades mecánicas. *Rev. Iberoam. De Polímeros* **2014**, *15*, 275–285. Available online: https://dialnet.unirioja.es/servlet/articulo?codigo=4801443 (accessed on 5 June 2023).
21. Carchi Maurat, D.E. Aprovechamiento de los Residuos Agrícolas Provenientes del Cultivo de Banano para Obtener Nanocelulosa. Bachelor's Thesis, University of Cuenca, Cuenca, Ecuador, 2014.
22. Siqueira, G.; Bras, J.; Dufresne, A. Cellulose whiskers versus microfibrils: Influence of the nature of the nanoparticle and its surface functionalization on the thermal and mechanical properties of nanocomposites. *Biomacromolecules* **2009**, *10*, 425–432. [CrossRef]
23. Hubbe, M.A.; Rojas, O.J.; Lucia, L.A.; Sain, M. Cellulosic nanocomposites: A review. *Bioresources* **2008**, *3*, 929–980.
24. Saïd Azizi Samir, M.A.; Alloin, F.; Paillet, M.; Dufresne, A. Tangling effect in fibrillated cellulose reinforced nanocomposites. *Macromolecules* **2004**, *37*, 4313–4316. [CrossRef]

25. Koyama, S.; Haniu, H.; Osaka, K.; Koyama, H.; Kuroiwa, N.; Endo, M.; Hayashi, T. Medical Application of Carbon-Nanotube-Filled Nanocomposites: The Microcatheter. *Small* **2006**, *2*, 1406–1411. [CrossRef]
26. Siró, I.; Plackett, D. Microfibrillated cellulose and new nanocomposite materials: A review. *Cellulose* **2010**, *17*, 459–494. [CrossRef]
27. Gardner, D.J.; Oporto, G.S.; Mills, R.; Samir, M.A.S.A. Adhesion and surface issues in cellulose and nanocellulose. *J. Adhes. Sci. Technol.* **2008**, *22*, 545–567. [CrossRef]
28. de Santayana, M.D.C.P. Nanotecnología y Alimentación. Ph.D. Thesis, Universidad Complutense. 2018. Available online: http://147.96.70.122/Web/TFG/TFG/Memoria/MARIA%20DEL%20CARMEN%20PARDO%20DE%20SANTAYANA%20DE%20PABLO.pdf (accessed on 21 June 2023).
29. Méndez, N.K.C. Tendencias investigativas de la nanotecnología en empaques y envases para alimentos. *Rev. Lasallista De Investig.* **2014**, *11*, 18–28.
30. RIKILT and JRC. Inventory of Nanotechnology Applications in the Agricultural, Feed and Food Sector. *EFSA Supporting Publication*. 2014. Available online: https://www.efsa.europa.eu/en/supporting/pub/en-621 (accessed on 12 May 2023).
31. Cozmuta, A.M.; Peter, A.; Cozmuta, L.M.; Nicula, C.; Crisan, L.; Baia, L. Active packaging system based on Ag/TiO$_2$ nanocomposite used for extending the shelf life of bread. *Packag. Technol. Sci.* **2015**, *28*, 271–284. [CrossRef]
32. Rosales, R.P.; Rodríguez, S.V.; Ramírez, R.C. El aceite de aguacate y sus propiedades nutricionales. *e-Gnosis* **2005**, *3*, 1–11.
33. ANIAME. El aceite de Aguacate en México. *Rev. ANIAME* **2002**, *8*, 1–9.
34. Mataix, J.; Gil, A. *Lípidos Alimentarios. Libro Blanco de los Omega-3. Los Ácidos Grasos Poliinsaturados Omega-3 y Monoinsaturados tipo Oleico y su Papel en la Salud*; Editorial Puleva; Instituto Omega: Granada, Spain, 2002; pp. 13–33.
35. Romero, R.A.; Bustamante, A.H.; Dávila, F.R.; Rodríguez, C.J.; Sánchez, N.A.; Rouzaud, S.O.; Canizales, R.D.F.; Otero, L.C.B.; Sánchez, M.R.I. Elaboración de sucedáneo saludable de mayonesa a base de aguacate (*Persea americana*) utilizando aislado de proteína de soya (*Glycine max*) como emulsificante. *Investig. Desarro. Cienc. Tecnol. Aliment.* **2016**, *1*, 591–597.
36. Durán Paz, S. Desarrollo de emulsiones a base de agentes antifúngicos con aceites esenciales, para reducir la incidencia del Stem-End Rot en frutas: Evaluación en aguacate hass. Bachelor's Thesis, Universidad de los Andes, Cundinamarca, Colombia, 2020.
37. Echeverri Romero, A.; Flórez Bulla, V. Elaboración y caracterización de una salsa vegana tipo mayonesa a base de aceite de aguacate Hass y aceite de semilla de Sacha Inchi. Bachelor's Thesis, Universidad de los Andes, Cundinamarca, Colombia, 2022.
38. Pathakoti, K.; Manubolu, M.; Hwang, H. Nanostructures: Current uses and future applications in food science. *J. Food Drug Anal.* **2017**, *25*, 245–253. [CrossRef]
39. EFSA Panel on Food Additives and Nutrient Sources added to Food (ANS); Mortensen, A.; Aguilar, F.; Crebelli, R.; Di Domenico, A.; Dusemund, B.; Frutos, M.J.; Galtier, P.; Gott, D.; Gundert-Remy, U.; et al. Re-evaluation of sorbitan monostearate (E 491), sorbitan tristearate (E 492), sorbitan monolaurate (E 493), sorbitan monooleate (E 494) and sorbitan monopalmitate (E 495) when used as food additives. *EFSA J.* **2017**, *15*, e04788. [CrossRef]
40. Roohani, M.; Habibi, Y.; Belgacem, N.M.; Ebrahim, G.; Karimi, A.N.; Dufresne, A. Cellulose whiskers reinforced polyvinyl alcohol copolymers nanocomposites. *Eur. Polym. J.* **2008**, *44*, 2489–2498. [CrossRef]
41. Cao, X.; Xu, C.; Wang, Y.; Liu, Y.; Liu, Y.; Chen, Y. New nanocomposite materials reinforced with cellulose nanocrystals in nitrile rubber. *Polym. Test.* **2013**, *32*, 819–826. [CrossRef]
42. Ilyas, R.A.; Sapuan, S.M. Biopolymers and Biocomposites: Chemistry and Technology. *Curr. Anal. Chem.* **2020**, *16*, 500–503. [CrossRef]
43. Mahendra, I.P.; Wirjosentono, B.; Tamrin, T.; Ismail, H.; Mendez, J.A.; Causin, V. The effect of nanocrystalline cellulose and TEMPO-oxidized nanocellulose on the compatibility of polypropylene/cyclic natural rubber blends. *J. Thermoplast. Compos. Mater.* **2022**, *35*, 2146–2161. [CrossRef]
44. Jain, M.; Pradhan, M.K. Morphology and mechanical properties of sisal fiber and nano cellulose green rubber composite: A comparative study. *Int. J. Plast. Technol.* **2016**, *20*, 378–400. [CrossRef]
45. Dasso, G. PLA/nanocelulosa: Nanobiomateriales para Envases. Bachelor's Thesis, Universidad Nacional de Mar de plata, Mar del Plata, Argentina, 2017.
46. Pech-Cohuo, S.C.; Canche-Escamilla, G.; Valadez-González, A.; Fernández-Escamilla, V.V.; Uribe-Calderon, J. Production and Modification of Cellulose Nanocrystals from *Agave tequilana* Weber Waste and Its Effect on the Melt Rheology of PLA. *Int. J. Polym. Sci.* **2018**, *2018*, 3567901. [CrossRef]

Disclaimer/Publisher's Note: The statements, opinions and data contained in all publications are solely those of the individual author(s) and contributor(s) and not of MDPI and/or the editor(s). MDPI and/or the editor(s) disclaim responsibility for any injury to people or property resulting from any ideas, methods, instructions or products referred to in the content.

Proceeding Paper

Study of the Mechanical Properties of Gels Formulated with Pectin from Orange Peel [†]

Nicolás Mauricio Bogdanoff [1,2,*], Camilo J. Orrabalis [3], Wilson Daniel Caicedo Chacon [4] and Germán Ayala Valencia [4]

1. Departament of Industrial Engineering, Facultad de Ingeniería, Universidad Nacional de Asunción, Campus Universitario de San Lorenzo, San Lorenzo 11001-3291, Paraguay
2. Centro de Investigaciones y Transferencia Formosa CONICET, Formosa P3600, Argentina
3. Laboratorio de Ingeniería de Materiales y Nanotecnología, Universidad Nacional de Formosa, Formosa P3600, Argentina; javi_c_o@hotmail.com
4. Department of Chemical and Food Engineering, Federal University of Santa Catarina, Florianópolis 88040-970, Brazil; w.caicedo.ch@gmail.com (W.D.C.C.); g.ayala.valencia@ufsc.br (G.A.V.)
* Correspondence: nbogdanoff@fiuna.edu.py; Tel.: +54-3454109336
† Presented at the 1st International Conference of the Red CYTED ENVABIO100 "Obtaining 100% Natural Biodegradable Films for the Food Industry", San Lorenzo, Paraguay, 14–16 November 2022.

Abstract: Pectin is a polysaccharide that is known for its gelling properties and its applications in the pharmaceutical industry. This can be divided into two structural groups, high methoxyl pectins (HMP) and low methoxyl pectins (LMP). Currently, there is little information on the properties of the orange pectin in which LPM predominates. The aim of this study was therefore to investigate the mechanical properties of gels produced with pectin isolated from orange peels. The results showed similar values to those found in the literature, except for hardness. The gels produced from the pectin could be used in the industry, the formulation varying depending on the application.

Keywords: low methoxyl pectins; amidated pectins; pharmaceutical industry; gelling property

1. Introduction

Pectin is a heteropolysaccharide found in plant cell walls and is known for its gelling properties and applications in the pharmaceutical industry. The power of pectin lies in the fact that it may strongly modify the structure of a solution to generate a gelled network, as well as the fact that it is of natural origin and has numerous healthy properties, which has resulted in its increased use for the formulation of edible gels. Indeed, the combination of characteristics such as vegetable origin, functionality, safety at high concentrations, commercial availability for a wide variety of products, and ease of production and application, are some advantages of pectins over other gelling agents [1,2]. The gelling property of an edible product can be beneficial in many ways. Some typical examples can be as simple as the pleasure and relaxing texture of a smooth, gelled dessert [3]. On the other hand, gels have other technical applications, such as the intake of bitter drugs, or the in situ release of drugs for specific pharmaceutical applications [4]. In the food industry, pectins have been used in a wide variety of products including beverages, confectionery, bakery, dairy, and meat.

Recall that the central molecule of pectin is a linear chain of α(1,4)-D-galacturonic acid, occasionally interrupted by (1,2)-L-rhamnose residues. Pectins can be divided into two structural groups: high methoxyl pectins (HMPs) with a degree of esterification (DE) (or methoxylation/methylation (DM)) higher than 50%, and low methoxyl pectins (LMPs) with a DE lower than 50%. Some carboxyl groups of galacturonic acid can be substituted with amidated groups. This class of pectins are called amidated low methoxyl pectins (ALMPs) and are characterized by their degree of amidation (DA). Intrinsic factors such as DE and

DA [5], the degree of polymerization (DP), and methoxylation patterns are key parameters that affect the behavior of pectins. In addition, extrinsic factors such as pectin and calcium concentration, pH, temperature, total soluble solids, different types of sugars, and metal ions, significantly impact the characteristics of a pectin-based gel. Many interactions can be anticipated when pectin molecules are used in product formulation with other molecules, such as carbohydrates and proteins. HMPs, LMPs, and ALMPs have different gelation mechanisms. According to Singhal [6], the orange has low methoxyl pectins when they are extracted at 100 °C.

The aim of this study was investigating the mechanical properties of gels produced with pectin isolated from orange peels.

2. Materials and Methods

2.1. Pectin Extract

To obtain the extract, the method of Canteri-Schemin et al. [7], with some modifications was used. The pectin was extracted with a solution of citric acid (Brand: Anedra, chemically pure) with a pH = 2.3 (measured using a Boeco pH meter, model BT500), and distilled water was used as a solvent. The process was carried out in flasks under reflux, with condensation at boiling temperature. The flasks were heated at 100 °C by electric heating mantles. The peel was added to the cold solvent at a peel/solvent ratio (Rs) of five. The initial time was considered when the solvent boiled, the total time for the process was 60 min. Agitation was achieved by the solvent moving on its own due to the boiling state. Once the processing time was completed, the extract from the exhausted peel was separated by means of a cloth filter. Filtration was carried out while hot.

Samples for analysis were taken immediately after filtering due to the low microbiological stability of the filter.

2.2. Physicochemical Determinations

2.2.1. Degree of Esterification

The degree of esterification of the pectin was determined according to the Dominiak technique [8]. Pectin samples are washed in a 60% 2-propan-ol solution containing 5% HCl, then washed with a 60% and 10% 2-propan-ol solution. Next, 0.2 g of the washed and dried material was dissolved in 100 mL deionized water and the sample is titrated with a 0.1 M NaOH solution using phenolphthalein as an indicator (the volume of the 0.1 M NaOH solution is referred to as V_1). The sample was then saponified by adding 10 mL of 1 M NaOH solution, followed by stirring for 15 min. Subsequently, 10 mL of 1 M HCl was added, and the sample was titrated again with 0.1 M NaOH until the color changes (volume V_2). The degree of esterification DE was calculated according to Equation (1).

$$DE = \frac{V_2}{V_1 + V_2} \cdot 100 \qquad (1)$$

2.2.2. Alcohol Precipitation

The modified Ranganna [9] technique was used. The extract was mixed with 3 volumes of ethanol. It was stirred for 3 min and left to rest for 1 h. The precipitate was then separated using a cloth filter and dried to a constant weight in a vacuum oven (AHR 8601) at a temperature of 45 °C. Then, it was ground in a mill (Control Química S.A., Model MC-1). The sample was packaged in glass vials and stored in a desiccator.

2.2.3. Gel Preparation

The Ranganna [9] technique was used. Measure 425 mL of water into a previously tared beaker. Add 10 mL of the 6% sodium citrate solution and the 60% citric acid solution. The mixture was heated up to 80 °C with constant stirring. Low methoxyl pectin was mixed with 30 g of sugar and it was placed in the glass. When the mixture was warm, 25 mL of the calcium chloride solution was added. Then, the mixture was stored at 24–26 °C for 18–24 h in corresponding containers for texture measurements.

As observed in the technique, the amount of pectin and calcium chloride to be added were considered variables since they are the parameters being evaluated. The Brix degrees will always remain fixed at around 35% (those normally used in a low-calorie formulations).

2.2.4. Mechanical Properties

Texture profile analysis of pectin gels were carried out using a technique proposed by Rascón-Chú et al. [10]. The gels were formed in 6 mL glass beakers, and the TPA was obtained using a TA.XT2i texture analyzer (RHEO Stable Micro Systems, Surrey, UK). Gels were compressed at a constant speed of 1 mm/s up to a distance of 3 mm from the gel surface using a cylindrical tip of 20 mm diameter and a trigger force of 5 g.

The measurements were carried out at room temperature (25–28 °C).

2.3. Statistical Analysis

The effect of calcium and pectin concentrations was explained by Hoefler [11], which was taken into account to observe the effects in terms of the rheological behavior of the gel formed, evidencing that the maximum peaks of gel strength have been between approximately 20 and 40 mg of calcium for each gram of low methoxyl pectin.

Statistical process optimizations using RSM have been widely employed by a number of researchers [12,13]. The Box–Behnken design of RSM was used to investigate the effects of two different independent variables: pectin yield and calcium concentration. The levels of these variables were selected based on the work of Hoefler [11]. The experiments were performed in random order. ANOVA was also performed to assess whether there are significant differences between the different formulations.

The statistical design used is detailed in Table 1.

Table 1. Statistical design of gelling experiences.

Calcium Concentration (mg/L)	Pectin Concentration (% w/w)				
	2	2.5	3	3.5	4
20	1.1	1.2	1.3	1.4	1.5
25	2.1	2.2	2.3	2.4	2.5
30	3.1	3.2	3.3	3.4	3.5
35	4.1	4.2	4.3	4.4	4.5
40	5.1	5.2	5.3	5.4	5.5
45	6.1	6.2	6.3	6.4	6.5

Which yields a total of 30 different formulations. All samples were tested in triplicate with a coefficient of variation of less than 10%.

3. Results

After carrying out the analytical technique by quintuplicate to determine the degree of methoxylation, it was obtained that the extracted pectin has a degree of 32.5%, therefore it is considered to be low methoxyl (LMP), probably due to the intensity of the extraction treatment, consequently, gels must be prepared with added calcium.

Once the gels were prepared according to the technique described above, they were left to rest for 24 h at room temperature (24–28 °C). After that time, no appreciable syneresis (liquid loss) was observed.

Table 2 shows the results of the TPAs produced by the texturometer for the different gel formulations.

Hardness is the only parameter that shows significant differences ($p < 0.05$) between the various gel formulations; the other parameters showed non-significant differences ($p > 0.05$), and the p values were calculated using Minitab 17 software. The values are within the order found by Pancerz et al. [14], who obtained a TPA of apple pectin gels at concentrations of 1.5% and 3%. The hardness values found here are a little higher than those of Rascón-Chu [10], however, the rest of the parameters are quite close to those obtained by these authors. Also, Urias-Orona et al. [15] determined hardness values of 3% apple pectin gels, which was very similar to those obtained in the present study. For a better understanding of the influence of the pectin and calcium concentration parameters on the hardness of the gels, a response surface, Figure 1, and a contoured surface, Figure 2, were plotted using Minitab 17 software.

Table 2. TPA results of pectin gels.

Cod	Hardness (g)	Adhesiveness	Cohesiveness	Elasticity	Gumminess	Chewability
1.1	38.00	−294	0.30	0.80	8.00	1.2
1.2	38.27	−295	0.32	0.82	8.20	1.1
1.3	52.97	−293	0.39	0.79	8.00	1.0
1.4	56.91	−264	0.35	0.80	8.30	1.4
1.5	57.82	−270	0.42	0.85	8.25	1.3
2.1	58.20	−268	0.38	0.15	8.20	1.3
2.2	59.04	−270	0.37	0.81	8.10	1.4
2.3	79.09	−285	0.37	0.85	8.15	1.3
2.4	84.54	−292	0.38	0.84	8.05	1.2
2.5	89.35	−300	0.41	0.85	8.08	1.5
3.1	80.05	−270	0.38	0.85	8.20	1.5
3.2	87.40	−265	0.42	0.88	8.40	1.3
3.3	93.04	−274	0.37	0.83	8.10	1.4
3.4	102.50	−286	0.40	0.89	8.35	1.5
3.5	123.20	−295	0.41	0.86	8.45	1.4
4.1	78.40	−280	0.39	0.84	8.25	1.3
4.2	84.23	−282	0.41	0.84	8.30	1.4
4.3	90.10	−289	0.39	0.82	8.31	1.5
4.4	100.30	−265	0.42	0.87	8.40	1.5
4.5	119.32	−270	0.40	0.89	8.50	1.4
5.1	49.20	−265	0.43	0.75	8.60	1.4
5.2	54.45	−246	0.35	0.81	8.90	1.2
5.3	60.15	−285	0.41	0.76	8.40	1.3
5.4	63.50	−289	0.38	0.74	8.65	1.2
5.5	70.00	−295	0.40	0.83	8.60	1.1
6.1	43.60	−290	0.35	0.69	8.80	1.2
6.2	50.80	−295	0.40	0.72	8.90	1.3
6.3	53.50	−301	0.39	0.76	8.53	1.2
6.4	56.00	−273	0.43	0.74	8.64	1.2
6.5	58.14	−270	0.45	0.76	8.70	1.1

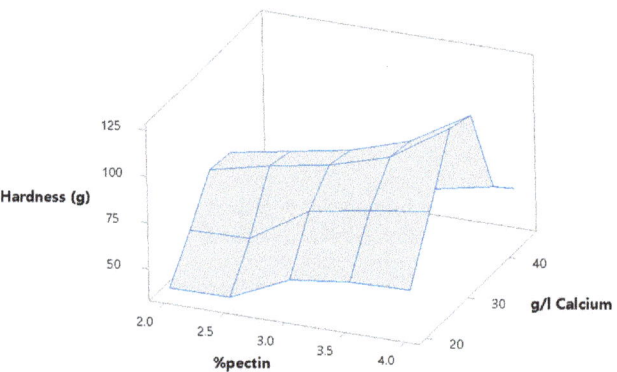

Figure 1. Response surface for hardness as a function of calcium concentration (g/L) and % pectin.

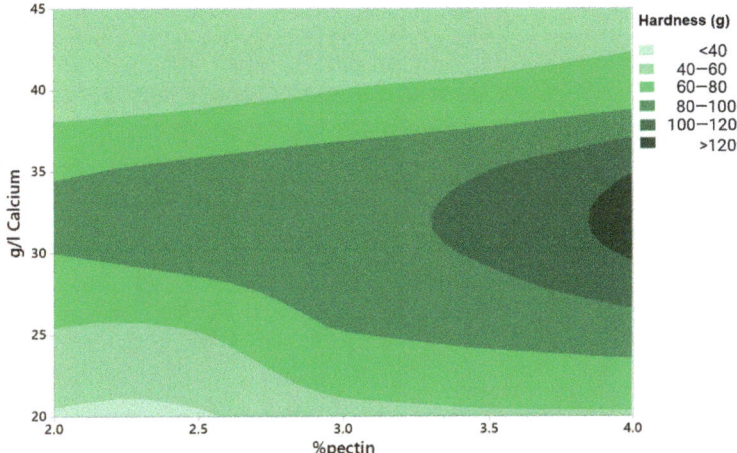

Figure 2. Contour surface for hardness as a function of calcium concentration (g/L) and % pectin.

4. Conclusions

The responses obtained coincide with what was anticipated in the literature. There was a region where the maximum hardness was obtained that corresponds to the interval of 30–35 mg/L of calcium. Naturally, as the pectin concentration increased, the calcium concentration was kept constant, and the hardness of the gel increased. The optimal formulation depends directly on the final use of the gel in the industry.

Author Contributions: N.M.B.: conceptualization, methodology, investigation, validation, formal analysis, writing—original draft preparation; C.J.O.: writing—review and editing, project administration, funding acquisition; W.D.C.C.: investigation supervision, writing—review and editing; G.A.V.: investigation supervision, writing—review and editing. All authors have read and agreed to the published version of the manuscript.

Funding: This research was funded by Red Cyted ENVABIO100 (Ref: 121RT0108), Universidad Nacional de Formosa (Disp. 011/20), Universidad Nacional de Entre Ríos (No. 8049)–Argentina.

Institutional Review Board Statement: Not applicable.

Informed Consent Statement: Not applicable.

Data Availability Statement: Not applicable.

Acknowledgments: The authors are grateful to the Rodolfo Mascheroni (UNLP), Oscar Iribarren (INGAR), Damián Stechina (UNER), and the Universidad Nacional de Entre Ríos.

Conflicts of Interest: The authors declare no conflict of interest.

References

1. Linares-García, J.A.; Ramos-Ramírez, E.G.; Salazar-Montoya, J.A. Viscoelastic properties and textural characterisation of high methoxyl pectin of hawthorn (*Crataegus pubescens*) in a gelling system. *Int. J. Food Sci. Technol.* **2015**, *50*, 1484–1493. [CrossRef]
2. Wang, J.; Aalaei, K.; Skibsted, L.H.; Ahrné, L.M. Bioaccessibility of calcium in freeze-dried yogurt based snacks. *LWT* **2020**, *129*, 109527. [CrossRef]
3. Rustagi, S. Food Texture and Its Perception, Acceptance and Evaluation. *Biosci. Biotechnol. Res. Asia* **2020**, *17*, 651–658. [CrossRef]
4. Rebitski, E.P.; Darder, M.; Carraro, R.; Ruiz-Hitzky, E. Chitosan and pectin core-shell beads encapsulating metformin-clay intercalation compounds for controlled delivery. *New J. Chem.* **2020**, *44*, 10102–10110. [CrossRef]
5. Belkheiri, A.; Forouhar, A.; Ursu, A.V.; Dubessay, P.; Pierre, G.; Delattre, C.; Djelveh, G.; Abdelkafi, S.; Hamdami, N.; Michaud, P. Extraction, characterization, and applications of pectins from plant by-products. *Appl. Sci.* **2021**, *11*, 6596. [CrossRef]
6. Singhal, S.; Hulle, N.R.S. Citrus pectins: Structural properties, extraction methods, modifications and applications in food systems—A review. *Appl. Food Res.* **2022**, *2*, 100215. [CrossRef]
7. Canteri-Schemin, M.H.; Cristina, H.; Fertonani, R.; Waszczynskyj, N.; Wosiacki, G. Extraction of Pectin from Apple Pomace. *Braz. Arch. Biol. Technol.* **2005**, *48*, 259–266. [CrossRef]

8. Dominiak, M.; Søndergaard, K.M.; Wichmann, J.; Vidal-Melgosa, S.; Willats, W.G.T.; Meyer, A.S.; Mikkelsen, J.D. Application of enzymes for efficient extraction, modification, and development of functional properties of lime pectin. *Food Hydrocoll.* **2014**, *40*, 273–282. [CrossRef]
9. Ranganna, S. *Handbook of Analysis and Quality Control for Fruit and Vegetable Products*, 1st ed.; Tata McGraw-Hill: New York, NY, USA, 1986.
10. Rascón-Chu, A.; Martínez-López, A.L.; Carvajal-Millán, E.; Ponce de León-Renova, N.E.; Márquez-Escalante, J.A.; Romo-Chacón, A. Pectin from low quality "Golden Delicious" apples: Composition and gelling capability. *Food Chem.* **2009**, *116*, 101–103. [CrossRef]
11. Hoefler, A.C. Other Pectin Food Products. In *The Chemistry and Technology of Pectin*, 1st ed.; Reginald, H.W., Ed.; Elsevier: Amsterdam, The Netherlands, 1991; Volume 1, pp. 51–66. [CrossRef]
12. Pishgar-Komleh, S.H. Application of Response Surface Methodology for Optimization of Picker-Husker Harvesting Losses in Corn Seed. *Iran. J. Energy Environ.* **2012**, *3*, 134–142. [CrossRef]
13. Asadi, A. Statistical Process Analysis and Optimization of an Aerobic SBR Treating an Industrial Estate Wastewater Using Response Surface Methodology (RSM). *Iran. J. Energy Environ.* **2011**, *2*, 356–365. [CrossRef]
14. Pancerz, M.; Kruk, J.; Ptaszek, A. The Effect of Pectin Branching on the Textural and Swelling Properties of Gel Beads Obtained during Continuous External Gelation Process. *Appl. Sci.* **2022**, *12*, 7171. [CrossRef]
15. Urias-Orona, V.; Rascón-Chu, A.; Lizardi-Mendoza, J.; Carvajal-Millán, E.; Gardea, A.A.; Ramírez-Wong, B. A novel pectin material: Extraction, characterization and gelling properties. *Int. J. Mol. Sci.* **2010**, *11*, 3686–3695. [CrossRef] [PubMed]

Disclaimer/Publisher's Note: The statements, opinions and data contained in all publications are solely those of the individual author(s) and contributor(s) and not of MDPI and/or the editor(s). MDPI and/or the editor(s) disclaim responsibility for any injury to people or property resulting from any ideas, methods, instructions or products referred to in the content.

Proceeding Paper

Preliminary Modeling Study of a Tape Casting System for Thermoplastic Starch Film Forming [†]

Liliana Ávila-Martín [1], Diana Katherine Guzmán Silva [1], Jairo E. Perilla [1,*] and Cristian Camilo Villa Zabala [2]

1 Departamento de Ingeniería Química y Ambiental, Universidad Nacional de Colombia 1, Bogotá 10839, Colombia; lavilam@unal.edu.co (L.Á.-M.); dguzmans@unal.edu.co (D.K.G.S.)
2 Grupo de Investigación en Ciencia y Tecnología de Alimentos-CYTA, Universidad del Quindío, Armenia 630001, Colombia; ccvilla@uniquindio.edu.co
* Correspondence: jeperillap@unal.edu.co
† Presented at the 1st International Conference of the Red CYTED ENVABIO100 "Obtaining 100% Natural Biodegradable Films for the Food Industry", San Lorenzo, Paraguay, 14–16 November 2022.

Abstract: Thermoplastic starch films (TPS) are an alternative for single-use plastics in packaging. Evaluating large-scale production alternatives that maintain the properties of these bio-based polymers is a crucial factor in understanding their potential industrial use. This preliminary study focuses on testing whether a mathematical model used to predict the drying conditions of ceramic film via tape casting can be adapted to the production of TPS. It also determines the possible drying tape speeds for this type of polymeric film.

Keywords: tape casting; film drying; starch

1. Introduction

Nowadays, a switch to environmentally friendly raw materials is increasingly necessary, which poses a new challenge for engineering since, in the industry, it becomes necessary to adapt manufacturing processes to the conditions tolerated by these new materials. Such is the case of bioplastics production [1,2], an industry in which some of the raw materials considered as possible substitutes require humidity and temperature conditions that are difficult to adapt in the polymer processing systems existing in the current plastics industry, or else they need pretreatments to acquire the characteristics that allow for their processability, which, at the same time, implies higher costs, such as in the case of thermoplastic starch or TPS [3,4].

There are two general processes for film manufacturing, the wet process or casting and the dry process. The first method consists of pouring a polymer solution onto a substrate and then drying the solvent, a technique also known as solvent casting, often used at the laboratory level. In addition, the second method is based on the thermoplastic properties presented by some biopolymers, such as the extrusion process [5]. Related to the dry method are also processes such as compression molding, injection molding, and blow molding; however, due to costs and characteristics of the final materials, the large-scale processing of TPS is not yet common worldwide, and research is still ongoing to improve both properties and large-scale manufacturing [6].

Therefore, considering that the decomposition temperature of starch (200–220 °C) is lower than the melting temperature (240 °C), which is why it cannot be processed as a polymer in its native state [4], the study of film properties of this biopolymer at the laboratory level generally begins with preparation via the manual casting molding technique, which allows for the use of starch gelatinized with water as a film-forming solution, which is poured into a mold that can be made of glass, acrylic, or Teflon. We would then wait for the solvent to evaporate, finally leaving a film in equilibrium with the humidity of the environment (dry) that can be removed from the mold (Figure 1). With this technique, film formation is favored, avoiding flow problems and the thermal degradation of biopolymer dispersions. Considering that the processing conditions required to transform biopolymers into thermoplastic materials via techniques such as extrusion can affect the properties of some of the components, especially the active ones [5,7], it is practical for experimentation; however, it is not useful in performing larger scale processes, as needed in the industry, since it does not allow for the preparation of film segments of large dimensions and requires long drying times [8]. In addition, some local irregularities are often unavoidable, such as variations in drying speed or final film thickness related to geometric and drying conditions. For these reasons, and due to the long drying time required, this methodology is not suitable for preparing larger films [9,10]. To solve this, solvent methods are used continuously, such as the tape casting method.

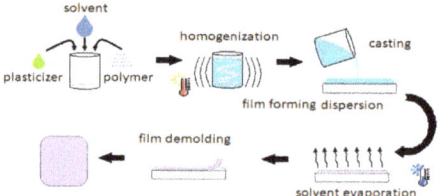

Figure 1. Solvent casting molding.

Although the tape system was developed in 1943 to produce ceramic films, this system has not changed much in recent years and consists of solution preparation equipment and a line of coating machines (Figure 2). The solution preparation equipment is a stand-alone batch operation and includes a thermostated vessel equipped with mixing and feeding systems; the latter consists of a coating applicator, which consists of a nozzle with a guide gate that is adjusted according to the final dry film thickness, usually set in micrometers; the nozzle, in turn, acts as a reservoir for the solution. On the other hand, there is the coating line on which the nozzle rests, which is a drag belt system pulled at a constant speed by a motor, which moves to cause the solution to be dragged between the guide gate and the conveyor belt, the result is a film in solution formed on the belt. This moves through the drying chamber that removes the solvent via evaporation, producing at the end of the process a dry film that is detached from the conveyor belt [5,7]. In recent years, the tape casting system has been considered an alternative for the continuous film production of biodegradable plastics, although there are still few studies on this topic, and despite allowing the formation of a continuous film, it is not common in the conventional plastics industry [6].

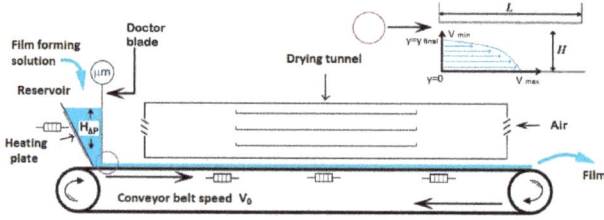

Figure 2. Tape casting system.

Studies have shown that a continuous casting method (blade coating or tape casting) can be used on an industrial scale for bioplastics production because the film-forming suspension is prepared on continuous conveyor belts with effective thickness control. The formed film is dried via heat conduction, convection, or radiation over short periods [11]. With this in mind, the wet method is a coating operation and, as such, pre-existing technologies can be adapted to the production of biopolymer films [5]. The production of edible films via dry methods means productivity and economics. However, considering the severe processing conditions required to transform biopolymers into thermoplastic materials, the properties of some of the components, especially the active ones, may be affected. Therefore, wet processes can offer a moderate and useful way to manufacture edible films, even account for the fact that it is an energy-intensive procedure [11].

When large-scale film preparation is required, it is essential to study the rheological properties and the effect of additives (plasticizers, fillers) on the thermoplastic behavior of film-forming materials in order to select the appropriate processing parameters [11]. Starch gels present thermoplastic behavior, which allows us to process films by applying different thermal and mechanical techniques. However, they present great changes in properties related to the change in starch concentration and its vegetable origin, as well as to the particular characteristics of the added additives. For this reason, specific analyses should be carried out for each type of filmogenic suspension to be processed; among the studies that have been reported for TPS, it is found that according to the rheological study of lecithin and leavening leucocytes, the following stand out: According to the rheological study of le Marcotte et al. (2001) [12], starch gels at different concentrations (4%, 5%, and 6%) present pseudoplastic behavior and absence of thixotropy. They were also reported to retard the sedimentation of solids added to the suspension, which facilitates the inclusion of fillers such as nanoparticles and fibers. On the other hand, differences in the proportion between amylose and amylopectin in starch generate changes in rheological properties, with gels with higher amylose content being more viscous [13], as well having as greater adhesiveness, in addition to presenting changes in thixotropy and pseudoplastic behavior [14]. Implying that a more detailed analysis of this type of gel is necessary when it is desired to scale up the production of films in systems such as tape casting, taking into account the flux pattern of the suspension is the main property related to the formation of good films via this technique [15].

For film formation, according to the needs of the industry, drying times must be taken into account. In addition, despite its frequent use in the food and chemical industry, reports on starch film drying are scarce in the literature; most of these reports refer to convection drying, and this absence is even more evident in studies related to the tape casting technique. We finding very few articles on this subject in the last decade, with most of them being carried out by a single research group.

The drying of starch films depends on several factors; for example, Oliveira de Moraes et al. (2015) [8] and Karapantsios (2006) [16] reported on the influence of temperature and suspension thickness on the conduction drying of starch-based films, considering that the thickness of the spread suspension was the most important variable controlling the properties of the films. De Moraes et al. (2013) used a doctor blade opening of 3-to-4 mm to obtain films with a maximum drying temperature of 60 °C. They found that suspensions of 3-mm thickness can be dried in 2.3 h. Furthermore, in de Moraes et al. (2013) [15], the authors report that the film thickness is always less than the blade spacing and depends on the spreading speed, using conveyor belt movement speeds of approximately 265 cm min and achieving drying times of 5 h, In both articles, shorter drying times are found than those reported in the literature for films prepared via solvent casting (more than 13 h) [17]. Other works, such as Mendes et al. (2020) [18], report that a 1.5 mm-thick layer that was conducted through two stages of oven drying at 90 °C allowed for complete drying. On the other hand, considering the form of heat transfer for film drying, a study by de Moraes and Laurindo (2018) [19] finds that infrared drying, compared to convection drying, as well as the coupling of these, can be a viable procedure for the large-scale production of starch-based films reinforced with cellulosic fibers. The article reports that drying with infrared radiation requires half the time of drying with heat conduction at 60 °C. However, higher infrared heating power may favor crack formation, affecting film properties.

On the other hand, only Elizabeth Gamboni et al. (2021) [5] evaluates the interaction between the conveyor belt material and the starch-based filmogenic solution as a factor in the design of a tape casting equipment. The study reports on the spreadability of the filmogenic solution and its adhesion ability, eventually choosing the polyurethane tape as the most suitable material for its performance and lower cost.

Finally, there is only one article that refers to the modeling of the biofilm production process via tape casting, that of Vogelsang et al. (2014) [7], which describes the preparation of a dextran-based film using water as solvent. The results are compared with a model in which correlations are established between the processing speed and the dry film thickness, which is the variable of interest at the end of the process. The flow is described by a nozzle, which, in its lower part, has a sheet that drags the fluid, forming a velocity profile in which the maximum velocity develops in the drag band, and the velocity in the nozzle is taken as zero assuming that there is no slip in that flow layer.

To develop the conical casting model for pseudoplastic fluids, it is described via the Navier–Stokes equation of motion (1) and the behavior of the viscous fluid with the power law (2), taking into account the solution proposed by Tadmor and Gogos (2006) [20].

$$\rho\left(\frac{\partial V_x}{\partial t} + V_x\frac{\partial V_x}{\partial x} + V_y\frac{\partial V_x}{\partial y} + V_z\frac{\partial V_x}{\partial z}\right) = -\frac{\partial P}{\partial x} - \left(\frac{\partial \tau_{xx}}{\partial x} + \frac{\partial \tau_{xy}}{\partial y} + \frac{\partial \tau_{xy}}{\partial z}\right) + \rho g_x \quad (1)$$

$$\tau_{xy} = k\left(\frac{dv_x}{dy}\right)^n \quad (2)$$

On the other hand, in the work of Tok et al. (2000) [21], a model is used for the processing of ceramics via tape casting, finding the thickness ratio in relation to the conveyor belt speed by performing the non-Newtonian combination of pressure and drag flow model. In this work, the power law equation is also used as a basis for describing the suspended flow, continuing with the coordinates of the Figure 2, where an infinite plane, $x - z$, located at $y = H_0/2$, which presents a maximum fluid velocity $Umax$, is assumed. No-slip conditions are considered, in which the same velocity of motion is presented for both the fluid and the plane. In the above conditions, the fluid presents a static pressure gradient (ΔP). Looking at the way the fluid moves in the reservoir nozzle and on the moving conveyor belt, it is observed that the model in the tape casting system can be described as a flow through parallel plates via Equation (3), where ρ_s and ρ_{tp} are, respectively, the density of the fluid and dry film; L is the channel length; and m and n the power law parameters. The correlation correction factor α is introduced for the reduction of the film extension at the nozzle outlet. In addition, because the thickness between the suspension and the film changes due to the effect of solvent evaporation, the factor β, which considers the correction for mass loss, must also be introduced. Thus, finally, the film thickness corresponds to

$$\delta_{tp} = \left[\frac{2\left(\frac{H_0}{2}\right)^{\frac{1}{n}+2}(\Delta P)^{\frac{1}{n}}}{L\left(\frac{1}{n}+2\right)m^{\frac{1}{n}}U} + \frac{1}{2}(H_0)\right]\frac{\alpha\beta\rho_s}{\rho_{tp}}. \quad (3)$$

Therefore, this preliminary study proposes the modeling of the conditions necessary in order to obtain a given film thickness as a function of the conveyor belt speed in a tape casting system for the production of starch-based bioplastic films, using the model proposed by Tok et al. (2000) for ceramics.

2. Methods

The model of Tok et al. (2000) [21] (Equation (3)) is used to predict the velocities required for starches. To compare the functionality of the model in biopolymers, the results reported by Vogelsang et al. (2014) [7] were used, obtaining the data from the graph reported in the article using Engauge Digitizer software. Velocity increments between 0 to 13 mm/s were used, and the pressure drop in the reservoir (756.25 Pa) was previously determined, using Excel solver to determine the film thickness parameters.

Finally, using the information found for cassava starch with fiber addition reported in Oliveira de Moraes et al. (2015) [8], outlined in Table 1, parameters are proposed for the calculation of speed and nozzle opening conditions in a tape casting system described by Vogelsang et al. (2014) [7] to prepare starch-based film. The values of m (3.545 Pa·s^n) and n (0.604), found in de Moraes et al. (2013) [15] for a temperature of 60 °C, were considered.

Table 1. Experimental data cassava starch film (Oliveira de Moraes et al., 2015) [8]. Temperature: 60 °C; conveyor belt speed: 41.67 mm/s.

Gap (m)	Density (Kg/m^3)	Drying Time (min)	Final Thickness (mm)
0.001	498	36	0.06
0.002	828	52	0.069
0.003	1025	67	0.118
0.004	1190	100	0.148

3. Results and Discussion

3.1. Validation

The experimental data, as well as those in the model, show a lower thickness with higher velocities, assuming then that the speed of the conveyor belt exceeds the flow velocity through the nozzle, which reduces the volume on the belt and, as a consequence, generates a lower thickness; then, the drag flow component becomes the dominant component at high speed. It is found that with the model of Tok et al. (2000) [21], the results found by Vogelsang et al. (2014) [7] can be predicted (Figure 3).

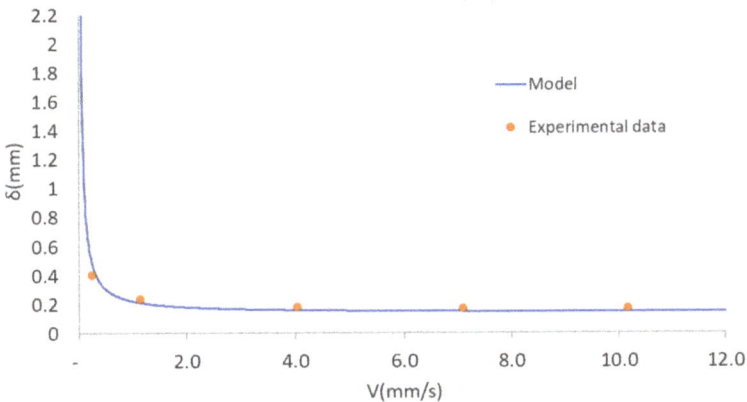

Figure 3. Film thickness vs. conveyor belt speed. Comparison between experimental results obtained by Vogelsang et al. (2014) [7] and theoretical results replicating the model proposed by Tok et al. (2000) [20].

3.2. Prediction of Starch Film Conditions

According to the graph constructed using the model of Tok et al. (2000) [21] (Figure 4), it is inferred that the larger the nozzle opening, the higher the speed required to obtain a thinner film. On the other hand, due to the change in the geometry of the equipment with which the calculations were performed, differences are found with the final thicknesses reported by Oliveira de Moraes et al. (2015) [8]. It is also expected that other conditions, such as drying time, ambient humidity, and length of the conveyor belt, among others, influence the results.

According to the results of the graph, it is expected that a nozzle opening of 3 mm is the most suitable for a starch using a speed of approximately 40 mm/s since it is the one that presents an average thickness, which could be considered acceptable.

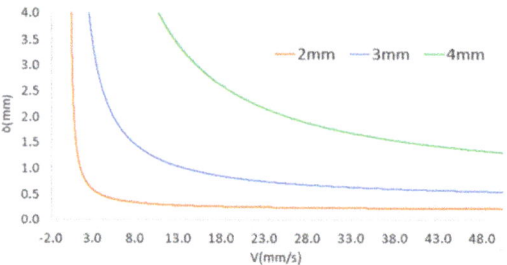

Figure 4. Prediction of cassava starch film thickness vs. conveyor belt speed with different nozzle openings (2 mm, 3 mm, and 4 mm).

4. Conclusions

Models proposed for ceramic materials processed via tape casting can be adapted to bioplastics with the same process.

The modeling of polymer processing via tape casting needs further study; it could present an opportunity to work with materials that require high moisture contents.

Author Contributions: Conceptualization, L.Á.-M. and J.E.P.; methodology, L.Á.-M. and J.E.P.; software, L.Á.-M. and D.K.G.S.; validation, L.Á.-M. and J.E.P.; formal analysis, L.Á.-M., D.K.G.S. and J.E.P.; investigation, L.Á.-M.; resources, L.Á.-M. and J.E.P.; data curation, L.Á.-M.; writing—original draft preparation, L.Á.-M. and J.E.P.; writing—review and editing, L.Á.-M. and J.E.P.; visualization, J.E.P.; supervision, J.E.P. and C.C.V.Z.; project administration, J.E.P.; funding acquisition, C.C.V.Z. and J.E.P. All authors have read and agreed to the published version of the manuscript.

Funding: This project was supported by grant "Program Becas de Excelencia Doctoral del Bicentenario—Cohorte I" and had funding from RED CYTED ENVABIO 100 (Ref: 121RT0108).

Institutional Review Board Statement: Not applicable for studies not involving humans or animals.

Informed Consent Statement: Not applicable for studies not involving humans.

Data Availability Statement: Not applicable.

Acknowledgments: RED CYTED ENVABIO 100 (Ref: 121RT0108) (interaction and publication cost).

Conflicts of Interest: The authors declare no conflict of interest.

References

1. Bátori, V.; Åkesson, D.; Zamani, A.; Taherzadeh, M.J.; Sárvári Horváth, I. Anaerobic degradation of bioplastics: A review. *Waste Manag.* **2018**, *80*, 406–413. [CrossRef] [PubMed]
2. Di Bartolo, A.; Infurna, G.; Dintcheva, N.T. A review of bioplastics and their adoption in the circular economy. *Polymers* **2021**, *13*, 1229. [CrossRef] [PubMed]
3. Reichert, C.L.; Bugnicourt, E.; Coltelli, M.B.; Cinelli, P.; Lazzeri, A.; Canesi, I.; Braca, F.; Martínez, B.M.; Alonso, R.; Agostinis, L.; et al. Bio-based packaging: Materials, modifications, industrial applications and sustainability. *Polymers* **2020**, *12*, 1558. [CrossRef] [PubMed]
4. Thomas, J. A Methodological Outlook on Bioplastics from Renewable Resources. *Open J. Polym. Chem.* **2020**, *10*, 21–47. [CrossRef]
5. Elizabeth Gamboni, J.; Verónica Colodro, M.; Marcelo Slavutsky, A.; Alejandra Bertuzzi, M. Selection of the conveyor belt material for edible film production by a continuous casting process. *Braz. J. Food Technol.* **2021**, *24*, e2020026. [CrossRef]
6. Do Val Siqueira, L.; Arias, C.I.; Maniglia, B.C.; Tadini, C.C. Starch-based biodegradable plastics: Methods of production, challenges and future perspectives. *Curr. Opin. Food Sci.* **2021**, *38*, 122–130. [CrossRef]
7. Vogelsang, D.F.; Perilla, J.E.; Buitrago, G.; Algecira, N.A. Preparation of biopolymer films by aqueous tape casting processing. *J. Plast. Film Sheeting* **2014**, *30*, 435–448. [CrossRef]
8. Oliveira de Moraes, J.; Scheibe, A.S.; Augusto, B.; Carciofi, M.; Laurindo, J.B. Conductive drying of starch-fiber films prepared by tape casting: Drying rates and film properties. *LWT* **2015**, *64*, 356–366. [CrossRef]
9. Bertuzzi, M.A.; Slavutsky, A.M. Standard and New Processing Techniques Used in the Preparation of Films and Coatings at the Lab Level and Scale-Up. In *Edible Films and Coatings: Fundamentals and Applications*; CRC Press: Boca Raton, FL, USA, 2016; pp. 21–42.

10. De Azeredo, H.M.; Rosa, M.F.; De Sá, M.; Souza Filho, M.; Waldron, K.W. The use of biomass for packaging films and coatings. In *Advances in Biorefineries: Biomass and Waste Supply Chain Exploitation*; Elsevier Ltd.: Amsterdam, The Netherlands, 2014; pp. 819–874.
11. Mellinas, C.; Valdés, A.; Ramos, M.; Burgos, N.; del Carmen Garrigós, M.; Jiménez, A. Active edible films: Current state and future trends. *J. Appl. Polym. Sci.* **2016**, *133*. [CrossRef]
12. Marcotte, M.; Taherian Hoshahili, A.R.; Ramaswamy, H.S. Rheological properties of selected hydrocolloids as a function of concentration and temperature. *Food Res. Int.* **2001**, *34*, 695–703. [CrossRef]
13. Xie, F.; Yu, L.; Su, B.; Liu, P.; Wang, J.; Liu, H.; Chen, L. Rheological properties of starches with different amylose/amylopectin ratios. *J. Cereal Sci.* **2009**, *49*, 371–377. [CrossRef]
14. Liu, Y.; Chen, X.; Xu, Y.; Xu, Z.; Li, H.; Sui, Z.; Corke, H. Gel texture and rheological properties of normal amylose and waxy potato starch blends with rice starches differing in amylose content. *Int. J. Food Sci. Technol.* **2021**, *56*, 1946–1958. [CrossRef]
15. De Moraes, J.O.; Scheibe, A.S.; Sereno, A.; Laurindo, J.B. Scale-up of the production of cassava starch based films using tape-casting. *J. Food Eng.* **2013**, *119*, 800–808. [CrossRef]
16. Karapantsios, T.D. Conductive drying kinetics of pregelatinized starch thin films. *J. Food Eng.* **2006**, *76*, 477–489. [CrossRef]
17. Li, C.; Xiang, F.; Wu, K.; Jiang, F.; Ni, X. Changes in microstructure and rheological properties of konjac glucomannan/zein blend film-forming solution during drying. *Carbohydr. Polym.* **2020**, *250*, 116840. [CrossRef] [PubMed]
18. Mendes, J.F.; Norcino, L.B.; Martins, H.H.A.; Manrich, A.; Otoni, C.G.; Carvalho, E.E.N.; Piccoli, R.H.; Oliveira, J.E.; Pinheiro, A.C.M.; Mattoso, L.H.C. Correlating emulsion characteristics with the properties of active starch films loaded with lemongrass essential oil. *Food Hydrocoll.* **2020**, *100*, 105428. [CrossRef]
19. De Moraes, J.O.; Laurindo, J.B. Properties of starch–cellulose fiber films produced by tape casting coupled with infrared radiation. *Dry. Technol.* **2018**, *36*, 830–840. [CrossRef]
20. Tadmor, Z.; Gogos, C.G. *Principles of Polymer Processing Second Edition*; John Wiley & Sons: Hoboken, NJ, USA, 2006.
21. Tok, A.I.Y.; Boey, F.Y.C.; Lam, Y.C. Non-Newtonian fluid flow model for ceramic tape casting. *Mater. Sci. Eng. A* **2000**, *280*, 282–288. [CrossRef]

Disclaimer/Publisher's Note: The statements, opinions and data contained in all publications are solely those of the individual author(s) and contributor(s) and not of MDPI and/or the editor(s). MDPI and/or the editor(s) disclaim responsibility for any injury to people or property resulting from any ideas, methods, instructions or products referred to in the content.

Proceeding Paper

Application of Cellulose-Based Film for Broccoli Packaging [†]

Erika Paulsen *, Sofía Barrios and Patricia Lema

Tecnologías Aplicadas a Procesos Alimentarios, Instituto de Ingeniería Química, Facultad de Ingeniería, Universidad de la República, Montevideo PC 11300, Uruguay; sbarrios@fing.edu.uy (S.B.); plema@fing.edu.uy (P.L.)

* Correspondence: erikap@fing.edu.uy

[†] Presented at the 1st International Conference of the Red CYTED ENVABIO100 "Obtaining 100% Natural Biodegradable Films for the Food Industry", San Lorenzo, Paraguay, 14–16 November 2022.

Abstract: Broccoli is a highly perishable vegetable with unique nutritional characteristics. Modified atmosphere packaging (MAP) has proven to be a successful technology to extend broccoli shelf-life. The main disadvantage of MAP is the extensive use of petrochemical-based films resulting in huge quantities of domestic plastic waste. In this study, suitability of a biodegradable cellulose-based film for broccoli florets packaging was evaluated, as an alternative to polypropylene film. Florets packaged in cellulose-based film showed a high mass loss and extremely low in-package O_2 concentrations, which made this material unsuitable for broccoli packaging application. Improved gas and water vapor barrier properties should be considered for biodegradable packages, in order to make their application for vegetable packaging feasible.

Keywords: cellulose-based film; packaging; storage; postharvest shelf-life; quality; *Brassica oleracea* var. *italica*

1. Introduction

Broccoli is a vegetable highly valued by modern consumers due to its health-promoting properties. However, since broccoli has a high respiration rate, it presents accelerated senescence during storage and a short shelf-life [1]. Modified atmosphere packaging (MAP) has proven to be a successful technology to preserve broccoli quality and extend its shelf-life [2]. MAP technology consists in packaging horticultural products in permeable films. Inside the package, a modified atmosphere, with decreased O_2 concentration and increased CO_2 concentration with respect to normal air, is generated with the interplay of product respiration and package permeability. This modified atmosphere slows down the product respiration rate, extending product shelf-life [3]. The main disadvantage of MAP is the extensive use of petrochemical-based films, resulting in huge quantities of domestic plastic waste. Replacing these films with bio-based and biodegradable materials could contribute to reducing the environmental impact of plastics [4]. In this line, in recent years, cellulose-based materials have been developed for their application in food packaging [5]. The aim of this study was to evaluate the suitability of a cellulose-based film for broccoli florets packaging, as an alternative to conventional polypropylene film.

2. Materials and Methods

2.1. Plant Material and Experimental Design

Broccoli heads (*Brassica oleracea* var. *italica* cv. Legacy) were cut into florets, washed, disinfected (NaClO, 100 ppm), dried, and packaged under a passive modified atmosphere. Approximately 100 g of broccoli florets was packaged in micro-perforated biaxially orientated polypropylene (PP) and cellulose-based (NatureFlex[TM] NVS23, Futamura Group, Cumbria, UK) (CB) bags. Broccoli packaged in macro-perforated polypropylene was used as a control. Bags were sealed using a Supervac GK105/1 packaging machine (Wien, Austria) with air injection. Samples were stored at 4 °C during 14 d. At preselected storage

times (0, 7, and 14 d), three packages (each package constituted an experimental unit) were sampled for each packaging film. Different packages were used at each sampling point. Samples were immediately evaluated and the following quality attributes were measured throughout shelf-life: headspace gas composition, mass loss (ML), texture, and sensory attributes.

2.2. Headspace Gas Composition

O_2 and CO_2 concentration inside packages was measured using a gas analyzer (OXYBABY® 6.0, WITT-Gasetechnik, Witten, Germany), extracting a 6 mL sample directly from the package. Results were expressed as partial pressure (kPa) of O_2 and CO_2 inside the bags.

2.3. Mass Loss

Mass loss (ML) was calculated by weighing broccoli florets prior to packaging (day 0) and at each sampling point. It was expressed as a percentage of initial weight (%).

2.4. Texture

A texture analysis was performed using a TA.XT2i Texture Analyzer (Stable Micro Systems Ltd., Godalming, UK). The Texture Analyzer was equipped with a 3 mm diameter cylinder probe in order to evaluate hardness of broccoli florets' stalks through a penetration test. Test conditions used for measurements were a 2.0 mm s^{-1} pre-test speed, 1.0 mm s^{-1} test speed, 5.0 mm s^{-1} post-test speed, and 5 mm penetration distance. Data of force (N) versus time (s) were registered using Texture Exponent Software (Version 3.2, Stable Micro Systems Ltd., Godalming, UK). The hardness value was determined as maximum force (N) registered in the force vs. time curves. Measurements were made on 4 broccoli stalks per experimental unit.

2.5. Sensory Evaluation

Overall appearance, color, and odor of broccoli florets were individually scored using a subjective scale from 1 to 5. A panel composed of seven members with sensory evaluation experience in vegetable quality was trained and carried out the evaluation. The rating scale for overall appearance was 5 = excellent, as freshly harvested; 4 = very good, minor defects; 3 = fair, moderate defects; 2 = poor, major defects; and 1 = very poor, inedible. In the case of odor, 5 = typical odor; 4 = slight off-odor; 3 = moderate off-odor; 2 = strong off-odor; and 1 = rot odor. In the case of color, 5 = dark green; 4 = green, yellow traces; 3 = light green, slightly yellow; 2 = light green, very yellow; and 1 = yellow. A score of 3 was considered as the limit of marketability and a score of 2 as the limit of edibility [6].

2.6. Statistical Analysis

Two-way ANOVA considering packaging condition, storage time, and their interaction was performed and when significant differences were observed, Tukey's test was applied ($p < 0.05$). Data are expressed as the mean ± standard error. XLSTAT (Statistical and data analysis solution, Lumivero, Denver, CO, USA) software was used for statistical analyses.

3. Results and Discussion

3.1. Headspace Gas Composition

PP samples showed a slight modification of internal package atmosphere ($p < 0.05$), reaching equilibrium O_2 and CO_2 concentration of 17.5 ± 0.7 and 3.6 ± 0.8 kPa, respectively.

CB samples showed a rapid change in headspace composition, reaching O_2 and CO_2 concentration of 2.4 ± 1.2 and 22.5 ± 1.6 kPa at day 3, respectively (Figure 1). According to the literature, CO_2 concentrations higher than 20 kPa could induce fermentative mechanisms, which would be detrimental to product quality. Over time, there was an excessive accumulation of CO_2 (40.4 ± 1.6 kPa at day 9) and depletion of O_2 (0.5 kPa at day 9), which makes cellulose-based film unsuitable for application in the packaging of high-respiration-rate products such as broccoli.

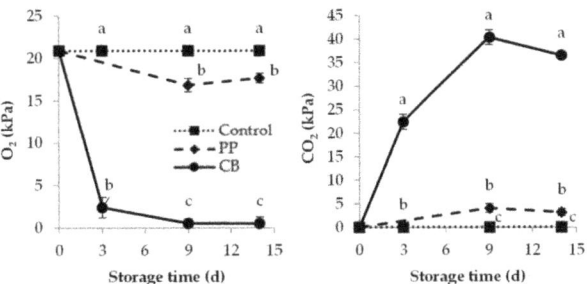

Figure 1. Effect of packaging condition on headspace O_2 and CO_2 concentration throughout storage at 4 °C. Mean values ($n = 3$) and standard error (vertical bars) are represented. Different letters indicate significant differences between packaging conditions at each sampling time ($p < 0.05$).

3.2. Mass Loss (ML)

Florets in cellulose-based film showed a marked ML, significantly higher than PP and control samples ($p < 0.0001$) (Figure 2). This result is striking because cellulose-based film provided a lower barrier to water vapor than the macro-perforated film. This could be explained with the high-water vapor transmission rate (WVTR) of the NatureFlex™ film, which causes a low relative humidity inside the package, thus increasing the broccoli transpiration rate. At day 9 of storage, CB samples showed ML values higher than 7%, which exceeds the marketability limit for fresh broccoli [7]. Therefore, the high WVTR of the cellulose-based film could be a limitation for its application on high-respiration- and -transpiration-rate products such as broccoli.

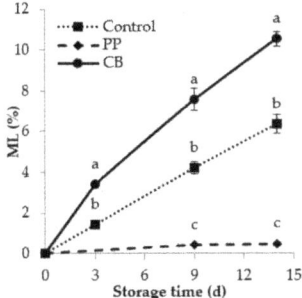

Figure 2. Effect of packaging condition on mass loss (ML) of broccoli florets throughout storage at 4 °C. Mean values ($n = 3$) and standard error (vertical bars) are represented. Different letters indicate significant differences between packaging conditions at the same sampling time ($p < 0.05$).

3.3. Texture

A significant effect of packaging film on broccoli florets' hardness was found (Figure 3). No change in hardness was observed for PP samples. CB samples showed a significant loss of hardness throughout storage ($p < 0.05$). This behavior could be due to the development of fermentative processes (low O_2 in-package concentration), which could damage the tissue structure [8], and to the extensive ML verified in these packages. Therefore, the cellulose-based film proved not to be a good packaging alternative for maintaining broccoli florets' texture.

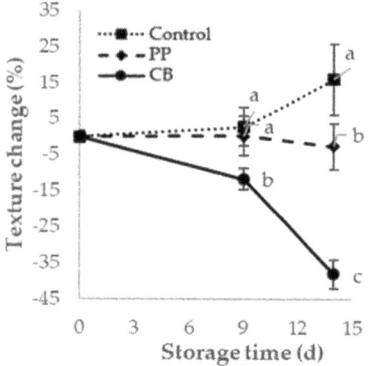

Figure 3. Effect of packaging condition on broccoli florets' hardness throughout storage at 4 °C. The data are expressed as the change in hardness relative to the initial value (%). Mean values ($n = 3$) and standard error (vertical bars) are represented. Different letters indicate significant differences between packaging conditions at the same sampling time ($p < 0.05$).

3.4. Sensory Evaluation

Sensory attributes' evolution of broccoli florets packaged in different films is shown in Figure 4. PP samples presented scores above the marketability limit throughout all of the storage period. Control florets showed a rapid and significant decrease in the color and overall appearance score. Florets in cellulose-based film showed no significant difference in the color score compared to PP florets. However, they showed a marked decrease in characteristic odor scores. Additionally, overall appearance decreased to the minimum score. Therefore, it can be assumed that gaseous conditions established inside the cellulose-based film favored the development of fermentative metabolism, producing volatile substances that generated off-odors, compromising broccoli florets' shelf-life.

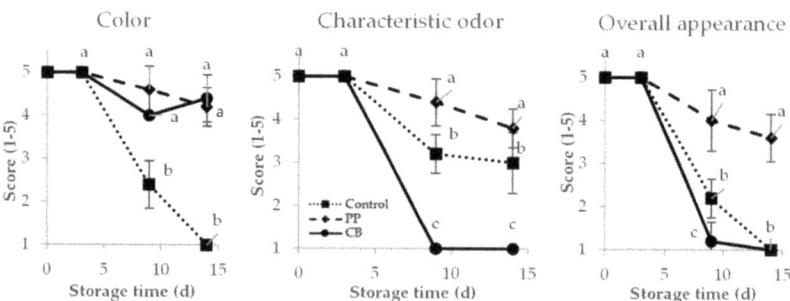

Figure 4. Effect of packaging condition on sensory attributes of broccoli florets throughout storage at 4 °C. Mean values ($n = 3$) and standard error (vertical bars) are represented. Different letters indicate significant differences between packaging conditions at the same sampling time ($p < 0.05$).

4. Conclusions

Too low O_2 concentrations and the excessive product mass loss presented with cellulose-based film make it unsuitable for packaging broccoli florets in the conditions assayed.

Interventions in the film that improve its gas and water vapor barrier properties should be considered to make its application in vegetable packaging feasible.

Author Contributions: Conceptualization, E.P. and P.L.; methodology, E.P.; software, E.P.; validation, S.B.; formal analysis, E.P.; investigation, E.P.; resources, P.L.; data curation, E.P.; writing—original draft preparation, E.P.; writing—review and editing, S.B.; visualization, E.P.; supervision, P.L.; project administration, P.L.; funding acquisition, P.L. All authors have read and agreed to the published version of the manuscript.

Funding: This research was funded by Comision Sectorial de Investigacion Científica, Universidad de la República and RED CYTED ENVABIO 100 (Ref: 121RT0108) (interaction and publication cost).

Institutional Review Board Statement: Not applicable.

Informed Consent Statement: Not applicable.

Data Availability Statement: Not applicable.

Acknowledgments: Authors are indebted to Agencia Nacional de Innovación e Investigación for granting Erika Paulsen a PhD scholarship (POS_EMHE_2018_1_1007740) and Red ENVABIO100-CYTED (Ref. 121RT0108).

Conflicts of Interest: The authors declare no conflict of interest.

References

1. Paulsen, E.; Barrios, S.; Baenas, N.; Moreno, D.A.; Heinzen, H.; Lema, P. Effect of temperature on glucosinolate content and shelf life of ready-to-eat broccoli florets packaged in passive modified atmosphere. *Postharvest Biol. Technol.* **2018**, *138*, 125–133. [CrossRef]
2. Jia, C.G.; Xu, C.J.; Wei, J.; Yuan, J.; Yuan, G.F.; Wang, B.L. Effect of modified atmosphere packaging on visual quality and glucosinolates of broccoli florets. *Food Chem.* **2009**, *114*, 28–37. [CrossRef]
3. Fonseca, S.C.; Oliveira, F.A.R.; Brecht, J.K. Modelling respiration rate of fresh fruits and vegetables for modified atmosphere packages: A review. *J. Food Eng.* **2002**, *52*, 99–119. [CrossRef]
4. Paulsen, E.; Lema, P.; Martínez-Romero, D.; García-Viguera, C. Use of PLA/PBAT stretch-cling film as an ecofriendly alternative for individual wrapping of broccoli heads. *Sci. Hortic.* **2022**, *304*, 111260. [CrossRef]
5. Shaikh, S.; Yaqoob, M.; Aggarwal, P. An overview of biodegradable packaging in food industry. *Curr. Res. Food Sci.* **2021**, *4*, 503–520. [CrossRef] [PubMed]
6. Winkler, S.; Faragher, J.; Franz, P.; Imsic, M.; Jones, R. Glucoraphanin and flavonoid levels remain stable during simulated transport and marketing of broccoli (*Brassica oleracea* var. *Italica*) heads. *Postharvest Biol. Technol.* **2007**, *43*, 89–94. [CrossRef]
7. Wiley, R.C. Preservation Methods for Minimally Processed Refrigerated Fruits and Vegetables. In *Minimally Processed Refrigerated Fruits & Vegetables*; Springer: Boston, MA, USA, 1994; pp. 66–134. Available online: https://link.springer.com/chapter/10.1007/978-1-4615-2393-2_3 (accessed on 1 October 2022).
8. Techavuthiporn, C.; Thammawong, M.; Nakano, K. Effect of short-term anoxia treatment on endogenous ethanol and postharvest responses of broccoli florets during storage at ambient temperature. *Sci. Hortic.* **2021**, *277*, 109813. [CrossRef]

Disclaimer/Publisher's Note: The statements, opinions and data contained in all publications are solely those of the individual author(s) and contributor(s) and not of MDPI and/or the editor(s). MDPI and/or the editor(s) disclaim responsibility for any injury to people or property resulting from any ideas, methods, instructions or products referred to in the content.

Proceeding Paper

Starch Nanoparticles Loaded with the Phenolic Compounds from Green Propolis Extract [†]

Maria Jaízia dos Santos Alves, Wilson Daniel Caicedo Chacon, Alcilene Rodrigues Monteiro and Germán Ayala Valencia *

Department of Chemical and Food Engineering, Federal University of Santa Catarina, Florianópolis 88040-970, Brazil; jaizia2011@gmail.com (M.J.d.S.A.); w.caicedo.ch@gmail.com (W.D.C.C.); alcilene.fritz@ufsc.br (A.R.M.)
* Correspondence: g.ayala.valencia@ufsc.br; Tel.: +55-48-3721-2534
[†] Presented at the 1st International Conference of the Red CYTED ENVABIO100 "Obtaining 100% Natural Biodegradable Films for the Food Industry", San Lorenzo, Paraguay, 14–16 November 2022.

Abstract: Phenolic compounds from propolis extract (PE) have antioxidant and antimicrobial properties; however, extracts from this raw material are not water soluble. This study aimed to stabilize the phenolic compounds from green propolis extract in cassava and potato starch nanoparticles produced by the anti-solvent precipitation method. The obtained materials displayed a crystalline structure related to starch nanomaterials with a V_{6h}-type crystalline structure. The starch nanoparticles interacted with the phenolic compounds by means of hydrogen bonds and increased the hydrophobicity in the nanomaterials. The developed starch nanomaterials loaded with the phenolic compounds from PE could be potentially used as a novel ingredient in food packaging.

Keywords: active nanomaterials; biopolymer; functional ingredients; food packaging

1. Introduction

Propolis is a resinous and heterogeneous material collected by *Apis mellifera* bees from different parts of plants, including the buds and exudates [1]. This natural compound has high amounts of flavonoids and phenolic acids, with it being used worldwide in traditional medicine due to its antioxidant and antimicrobial properties [2]. However, extracts from propolis have limited water solubility, reducing their application in the pharmacology and food industries [3].

Recently, Alves et al. [3] stabilized phenolic compounds from brown propolis extract using starch nanoparticles and observed that the obtained nanomaterials have high antioxidant activity. The authors concluded that the starch nanomaterials loaded with the phenolic compounds from brown propolis extract could be used as active ingredients in food packaging materials. Green propolis is another type of propolis abundant in Brazil, which has phenolic compounds with antioxidant and antimicrobial properties [4]. However, the stabilization of phenolic compounds from green propolis extracts using biopolymeric nanoparticles has not been investigated. Hence, this research aimed to produce and characterize cassava and potato starch nanoparticles loaded with the phenolic compounds from green propolis extract.

2. Materials and Methods

2.1. Materials

In the current research, starches isolated from cassava and potato were used as macromolecules. Native starches were purchased from Juréia and Shambala Naturais Food Industries (Florianópolis, Brazil). Green propolis was purchased from Breyer® (Formigas, Brazil). Distilled water, ethanol (≥99.6%, Êxodo Científica, São Paulo, Brazil), and hydrochloric acid (37 wt%, Neon, São Paulo, Brazil) were used as solvents. Potassium chloride and sodium carbonate were purchased from Dinâmica (São Paulo, Brazil). Folin–Ciocalteu reagent was acquired directly from Sigma-Aldrich (São Paulo, Brazil). All reagents used were of analytical grade, and they were used as received.

2.2. The Production of Starch Nanoparticles Loaded with the Phenolic Compounds from the Propolis Extract

Firstly, propolis extract (PE) was produced according to the methodology and best conditions described by Alves et al. [3]. In sequence, PE was acidified with hydrochloric acid (100:1 v/v, hydroethanolic solution: HCl 37 wt%, pH = 1). In parallel, starch nanoparticles (SNPs) were produced by the anti-solvent precipitation method [3,5]. Dispersions (5% w/w) of cassava starch and potato starch were prepared in distilled water at 25 °C followed by gelatinization at 90 °C for 30 min. The gelatinized starch solutions were cooled to 30 °C and then the acidified PE was dripped using a peristaltic pump (flow of 0.7 mL/min) in a 1:1 (% v/v) ratio.

The resulting slurry (starch dispersion + acidified PE) was stirred at 25 °C for 12 h and then centrifugated at 4000 rpm for 15 min using a centrifuge (Kasvi, São Paulo, Brazil). The SNPs were centrifuged three times with hydroethanolic solution (80% v/v) and finally washed with absolute ethanol (99.6%). The SNPs were separated by centrifugation and the ethanol was evaporated using a forced-air convection oven (Solidsteel, São Paulo, Brazil) at 60 °C for 10 min. Finally, the SNPs were frozen at −18 °C for 48 h and then lyophilized (Liotop L 101). The resulting nanomaterials loading the phenolic compounds from PE were named cassava (CSNPs-PE) and potato (PSNPs-PE) starch loading the phenolic compounds from PE, and cassava (CSNPs) and potato (PSNPs) starch nanoparticles without PE.

2.3. Characterization of the Starch Nanoparticles Loaded with the Phenolic Compounds from Green Propolis Extract

The loading efficiency (LE) of the total phenolic compounds (TPC) from the acidified PE stabilized with the SNPs was calculated using Equation (1) [6]:

$$LE(\%) = \frac{TPC_i - TPC_s}{TPC_i} * 100, \quad (1)$$

where TPC_i is the TPC of the PE and TPC_s is the TPC of supernatant collected after the first centrifugation. The quantification of TPC was carried out using the method described by Alves et al. [3].

Diffraction analysis was performed with an X-ray diffractometer (Rigaku MiniFlex 600 DRX, Tokyo, Japan) equipped with Cu-Kα radiation (λ = 0.154056 nm). XRD diffractograms were obtained 2θ = 3° and 60° (rate of 10 °/min). Equation (2) (Bragg's law) was used to calculate the interplanar spacing d (nm) from the X-ray patterns.

$$n\lambda = 2d\sin\theta, \quad (2)$$

where n is the reflection order (n = 1), λ is the wavelength of CuKα radiation, and θ is the reflection angle [7].

Chemical bonds were studied using a Fourier transform infrared spectrometer (FTIR, Cary 600, Agilent, Santa Clara, CA, USA) in the wavenumber range of 4000 and 400 cm^{-1} (4 cm^{-1} resolution). In each analysis, 32 scans were performed [7].

The water contact angle (WCA) of all starch nanoparticles was investigated using the methodology reported by Amirabadi et al. [8]. The samples were compressed into tablets using a hydraulic machine with two heated plates at 25 °C and controlled by PID controllers. A cylindrical piston was used as a mold. Approximately 0.2 g of each sample was deposited in the mold and 1 ton of force was applied. The pellets were approximately 1.5 mm thick and 1 cm in diameter. The water contact angle of the compressed samples was analyzed in an optical tensiometer (Ramé-Hart 250), with 5 µL of water being dropped over each compressed sample. The WCA was defined as the average of 10 measurements taken over a 5 s interval.

3. Results

Characterization of the Starch Nanoparticles Loaded with the Phenolic Compounds from Green Propolis Extract

The green propolis extract had a TPC of 763.36 mg GAE/g. After anti-solvent precipitation, the LE oscillated between 65.45 and 73.32% in PSNPs-PE and CSNPs-PE, respectively.

The samples revealed X-ray diffractograms of starch nanomaterials (Figure 1a). In particular, the X-ray diffractograms were typical of a V_{6h}-type crystalline structure, exhibiting diffraction peaks at 2θ = 13.0° (d = 0.68 nm) and 20.0° (d = 0.44 nm).

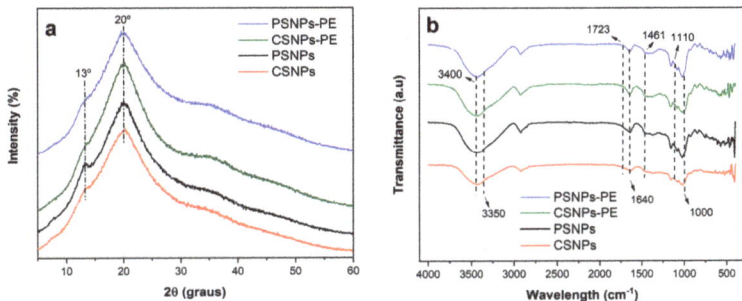

Figure 1. (**a**) X-ray diffractograms and (**b**) FTIR spectra of starch nanoparticles with (PSNPs-PE and CSNPs-PE) and without (PSNPs and CSNPs) PE.

The FTIR spectra of the samples show a peak centered at 3400 cm^{-1}, associated with the vibration of hydroxyl groups (O–H stretching) of the starch chains. A band at 3350 cm^{-1} was correlated with the O–H stretching vibration of the phenolic groups (Figure 1b) [9]. Furthermore, a slight band at 1723 cm^{-1} suggests the C=O stretching of the carboxylic group, indicating the presence of polyphenols from PE. Vibration of the phenol groups was also observed in the band centered at 1640 cm^{-1}, assigned to aromatic ring C=C stretching, as well as aromatic C–H deformation vibration at 1110 cm^{-1} [9]. C–H deformations and aromatic stretching at 1461 cm^{-1} was correlated with the presence of flavonoids (hydrocarbons CH_3 and CH_2's vibrations were overlapping) [9]. In the region around 1000 cm^{-1}, a new band was observed in starch nanoparticles loaded with the phenolic compounds from green PE.

The WCA of the CSNPs and PSNPs remained constant at 41.05° ± 0.17 (Figure 2). With the incorporation of the phenolic compounds from PE, an increase in the WCA was observed (Figure 2); hence, the starch nanoparticles loaded with the phenolic compounds from green PE had a WCA ranging between 66.80° ± 1.21 (CSNPs-PE) and 75.70° ± 0.75 (PSNPs-PE).

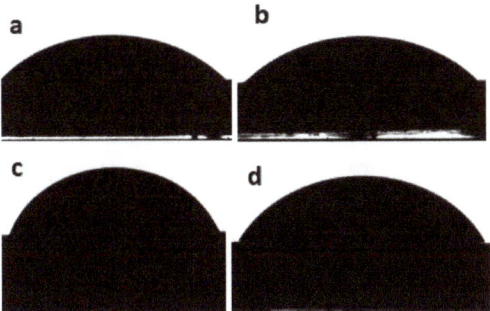

Figure 2. The WCA of potato (**a**) and cassava (**b**) starch nanoparticles and potato (**c**) and cassava (**d**) starch nanoparticles loaded with the phenolic compounds from green PE.

4. Discussion

In the current research, starch nanoparticles based on cassava and potato starches had similar LE values when compared with starch nanoparticles loaded with the phenolic compounds from brown propolis extract [3]. These results suggest that the LE could be independent of the type of propolis used in the PE.

Regarding the crystalline and chemical bond results, it is possible to conclude that the nanoparticles are composed of six glucose units per helical turn [10], with it being the case that the phenolic compounds altered this crystalline structure since a reduction in the peak intensity at 13° was observed in the X-ray diffractograms. Furthermore, the displacement observed at 1000 cm^{-1} in the FTIR spectra of the starch nanoparticles loaded with the phenolic compounds from PE suggests that structural modification resulted in spatial displacement of and an increase in the CO group band, probably caused by hydrogen bonds between the phenolic compound from PE and amylose/amylopectin chains [11].

Finally, the increase in the WCA values confirms the presence of phenolic compounds from PE in the starch nanoparticles. These phenolic compounds have hydrophobic properties and then increased the WCA values. The increase in the WCA could be important in packaging materials that will be used in contact with food.

5. Conclusions

In the current research, cassava and potato starch nanoparticles loaded with the phenolic compounds from propolis extract (PE) were produced and characterized. The developed nanomaterials displayed a V_{6h}-type crystalline structure, typical of starch nanoparticles. This crystalline structure was modified by the incorporation of phenolic compounds from PE. The FTIR results revealed that the starch chains interacted with the phenolic compounds from PE by means of hydrogen bonds. Finally, the starch nanoparticles had hydrophilic surfaces with a water contact angle (WCA) of 41.05°. With the incorporation of phenolic compounds from PE, the WCA in the starch nanomaterials increased between 60 and 80%, indicating that the phenolic compounds reduce the hydrophilicity of the nanoparticles. Based on these results, it can be considered that starch nanoparticles loaded with the phenolic compounds from PE can serve as promising ingredients to manufacture food packaging materials.

Author Contributions: M.J.d.S.A.: methodology, investigation, validation, formal analysis, and writing—original draft preparation; W.D.C.C.: investigation; A.R.M.: supervision and writing—review and editing; G.A.V.: conceptualization, formal analysis, resources, data curation, writing—original draft preparation, writing—review and editing, project administration, and funding acquisition. All authors have read and agreed to the published version of the manuscript.

Funding: This work was (partially) supported by Programa Iberomaricano de Ciencia y Tecnologia para el Desarrollo (CYTED) (through Red 121RT0108) and the Fundação de Amparo à Pesquisa e Inovação do Estado de Santa Catarina (FAPESC) (grants 2021TR000418 and 2021TR001887).

Institutional Review Board Statement: Not applicable.

Informed Consent Statement: Not applicable.

Data Availability Statement: Not applicable.

Acknowledgments: The authors gratefully acknowledge CAPES (Coordination for the Improvement of Higher Education Personnel) for the doctoral fellowship awarded to the first author and the Central Chemical Analysis of Chemical Engineering and Food Engineering for the analyses. G.A. Valencia would like to thank the CNPq (National Council for Scientific and Technological Development) for the research fellowship (302434/2022-4).

Conflicts of Interest: The authors declare no conflict of interest.

References

1. Bertotto, C.; Bilck, A.P.; Yamashita, F.; Anjos, O.; Bakar Siddique, M.A.; Harrison, S.M.; Brunton, N.P.; Carpes, S.T. Development of a biodegradable plastic film extruded with the addition of a Brazilian propolis by-product. *LWT* **2022**, *157*, 113124. [CrossRef]
2. Siripatrawan, U.; Vitchayakitti, W. Improving functional properties of chitosan films as active food packaging by incorporating with propolis. *Food Hydrocoll.* **2016**, *61*, 695–702. [CrossRef]
3. Jaízia dos Santos Alves, M.; Rodrigues Monteiro, A.; Ayala Valencia, G. Antioxidant Nanoparticles Based on Starch and the Phenolic Compounds from Propolis Extract: Production and Physicochemical Properties. *Starch/Stärke* **2022**, *74*, 2100289. [CrossRef]
4. Quintino, R.L.; Reis, A.C.; Fernandes, C.C.; Martins, C.H.G.; Colli, A.C.; Crotti, A.E.M.; Squarisi, I.S.; Ribeiro, A.B.; Tavares, D.C.; Miranda, M.L.D. Brazilian Green Propolis: Chemical Composition of Essential Oil and Their In Vitro Antioxidant, Antibacterial and Antiproliferative Activities. *Braz. Arch. Biol. Technol.* **2020**, *63*, e20190408. [CrossRef]
5. dos Santos Alves, M.J.; Calvo Torres de Freitas, P.M.; Monteiro, A.R.; Ayala Valencia, G. Impact of the Acidified Hydroethanolic Solution on the Physicochemical Properties of Starch Nanoparticles Produced by Anti-Solvent Precipitation. *Starch-Stärke* **2021**, *73*, 2100034. [CrossRef]
6. Quiroz, J.Q.; Velazquez, V.; Corrales-Garcia, L.L.; Torres, J.D.; Delgado, E.; Ciro, G.; Rojas, J. Use of plant proteins as microencapsulating agents of bioactive compounds extracted from annatto seeds (*Bixa orellana* L.). *Antioxidants* **2020**, *9*, 310. [CrossRef]
7. Capello, C.; Leandro, G.C.; Campos, C.E.M.; Hotza, D.; Carciofi, B.A.M.; Valencia, G.A. Adsorption and desorption of eggplant peel anthocyanins on a synthetic layered silicate. *J. Food Eng.* **2019**, *262*, 162–169. [CrossRef]
8. Amirabadi, S.; Milani, J.M.; Sohbatzadeh, F. Application of dielectric barrier discharge plasma to hydrophobically modification of gum arabic with enhanced surface properties. *Food Hydrocoll.* **2020**, *104*, 105724. [CrossRef]
9. Svečnjak, L.; Marijanović, Z.; Okińczyc, P.; Marek Kuś, P.; Jerković, I. Mediterranean Propolis from the Adriatic Sea Islands as a Source of Natural Antioxidants: Comprehensive Chemical Biodiversity Determined by GC-MS, FTIR-ATR, UHPLC-DAD-QqTOF-MS, DPPH and FRAP Assay. *Antioxidants* **2020**, *9*, 337. [CrossRef] [PubMed]
10. Shi, L.; Hopfer, H.; Ziegler, G.R.; Kong, L. Starch-menthol inclusion complex: Structure and release kinetics. *Food Hydrocoll.* **2019**, *97*, 105183. [CrossRef]
11. Pérez-Vergara, L.D.; Cifuentes, M.T.; Franco, A.P.; Pérez-Cervera, C.E.; Andrade-Pizarro, R.D. Development and characterization of edible films based on native cassava starch, beeswax, and propolis. *NFS J.* **2020**, *21*, 39–49. [CrossRef]

Disclaimer/Publisher's Note: The statements, opinions and data contained in all publications are solely those of the individual author(s) and contributor(s) and not of MDPI and/or the editor(s). MDPI and/or the editor(s) disclaim responsibility for any injury to people or property resulting from any ideas, methods, instructions or products referred to in the content.

Proceeding Paper

Microcrystals and Microfibers of Cellulose from *Acrocomia aculeata* (Arecaceae) Characterization [†]

Shirley Duarte [1,*], Magna Monteiro [2], Porfirio Andrés Campuzano [1], Natalia Giménez [1] and María Cristina Penayo [1,*]

1. Faculty of Chemistry, National University of Asunción, San Lorenzo 1055, Paraguay; campuandres@gmail.com (P.A.C.); ng76397@gmail.com (N.G.)
2. Polytechnic School, National University of Asuncion, Mcal. Estigarribia km 11, Asuncion 1209, Paraguay; mmonteiro@pol.una.py
* Correspondence: sduarte@qui.una.py (S.D.); mcpenayo@qui.una.py (M.C.P.)
† Presented at the 1st International Conference of the Red CYTED ENVABIO100 "Obtaining 100% Natural Biodegradable Films for the Food Industry", San Lorenzo, Paraguay, 14–16 November 2022.

Abstract: In the context of the so-called lignocellulose bio-refinery, the coconut shell (S) and pulp (P) of *Acrocomia aculeata* (Arecaceae) are interesting agro-industrial wastes that can be used as feedstock for the production of high value-added products. The aim of this work was to evaluate these lignocellulosic residues S and P, to obtain the microcrystal (MCC) and microfiber (MFC) of cellulose, and to characterize them to propose possible applications. First, cellulose content in the raw materials was determined, being 39.69% and 45.42% for both (S and P)) respectively, respectively. Then, the purification of residues was carried out via alkaline and bleaching treatments. Next, in order to obtain MCC and MFC from the purified cellulose, a chemical treatment with HCl (for MCC) and a mechanical treatment with a blender (for MFC) were performed. The size and morphology were observed via MEB, and properties were characterized using Fourier transform infrared spectroscopy (FTIR), X-ray diffraction (XRD), and differential thermogravimetric analysis (DTG).

Keywords: coconut fruit; shell; pulp; cellulose; microcrystals; microfiber

1. Introduction

The South American palm species *Acrocomia aculeata* (Arecaceae), commonly known as mbocayá, macaw, macauba, or just coconut palm, has attracted the attention of researchers in recent years, mainly for its great potential as a sustainable oil crop [1–3]. In Paraguay, the *A. aculeata* fruit (coconut) has been processed since 1940 [4] for oil extraction. From the process, the shell (S), pulp (P), endocarp, and almond expeller remain as waste, which can be used to obtain new products with greater added value [5,6].

Lignocellulosic wastes as S and P from coconut fruit represent a renewable cellulose source, the primary raw material for nano and micro celluloses (NCs and MCs) [7]. Cellulose derivatives with desirable properties for different applications are a current topic of study in the scientific community. Although NC is a material with exceptional properties, the high consumption of mineral acids during its extraction process is, in most cases, still a disadvantage.

Microcrystal (MCC) and microfiber cellulose (MFC) have gained increasing interest as reinforcing polymeric materials due to their availability, relatively low cost, and high mechanical resistance [8–10]. There are also numerous industrial applications for these fibers, which exploit their chemical functionality (reactivity) for crosslinking, their ability to retain water, and their hydrogen bonding capability [11].

MCC and MFC compositions must be studied for each source because characteristics such as crystallinity, thermal stability, and final chemical composition are dependent on their origin [12]. This information is essential to explore MFC and MCC uses as reinforcing agents in developing polymeric biodegradable compounds, e.g., packaging applications with improved properties.

This work describes the characteristics of two lignocellulosic agroindustrial wastes from coconut fruit (S and P), their main components, and the methods used to obtain MCC and MFC from them. The main characteristics of the MCC and MFCs, considering their possible application as reinforcing agents in films for food packaging, are also presented.

2. Materials and Methods

2.1. Raw Materials

Coconut shell (S) and pulp (P) were provided as agroindustrial waste by Industrial Aceitera S.A. paraguayan company. They were washed to remove dirt and other impurities. Then, they were dried in an oven at 105 ± 5 °C for 3 h. Finally, sample particles between 0.85 and 2 mm were obtained using a mill IKA M20, as NREL/TP51042620 establishes them.

2.2. Compositional Analysis

The chemical composition of the constituents (S and P) was determined: organic extractive (TAPPI 204 cm-97), holocellulose, and hemicelluloses based on the standard ASTM D1104 [13,14], soluble and insoluble lignin (NREL/TP510-42618).

2.3. Cellulose Preparation

Cellulose purification was carried out via two chemical treatments: (i) alkaline treatment with 5% NaOH (w/v) solution; (ii) bleaching process, using 1.5% NaClO2 solution (v/v) with acetic acid (pH 4–5), both treatments for 120 min at 80 °C, 600 rpm, and a solid/solution ratio of 1:20. The yield was calculated based on the dry basis weight of each constituent.

Microfibrillated and Microcrystal Cellulose Preparation

MFC was prepared by blending the bleached cellulose in a high rotation brand blender (model BL 767) at 25,000 rpm at different times (5, 10, 20 min), maintaining a ratio of 1% (p/v) (fiber/solution) in a volume of 500 mL of solution. At the end of the corresponding time of the mechanical treatment, the solution was filtered, and the MFCs obtained were dried in an oven at 45 °C for 18 h.

For the MCC, the bleached cellulose was subjected to different times of acid hydrolysis (15, 30, and 60 min) with 2.5 N HCl at a constant temperature of 85 ± 2 °C and a fiber/solution ratio (1:20), with a constant stirring. The reaction was stopped with an ice bath and adding NaOH. It was filtered through a fritted glass filter and washed with distilled water until pH neutral. Subsequently, it was dried in an oven at 45 ± 5 °C for around 18 h.

2.4. Characterization of MCC and MFC

2.4.1. Fourier Transform Infrared Spectroscopy (FTIR)

The presence of functional groups was studied using a Thermo Fisher (Nicolet iS5, Thermo Fisher, Waltham, MA, USA) FT/IR (Fourier transform infrared spectrometer). For the FTIR scan, KBr pellets containing 2 mg of dry sample with 200 mg of KBr powder were prepared. Spectra in the range of 400–4000 cm^{-1} was obtained with a resolution of 4 cm^{-1}, and the signal was accumulated from 32 scans [15,16].

2.4.2. X-ray Diffraction (XRD)

X-ray diffractometer was measured using the X'Pert^3 Powder via Cu-Kα radiation at 45 kV and 40 mA in an angular range of 10° to 40°/2θ with a step size of 0.0170/2θ and a count time of 50.1650 s in each step.

The crystallinity index (CrI) was measured using Segal's Equation on the diffraction peak height intensity, as indicated in Equation (1):

$$\%CrI = \frac{I_{002} - I_{am}}{I_{002}} \qquad (1)$$

where I_{002} is the maximum intensity of the diffraction peak, taken at 2θ between 22° and 23° for cellulose I, and I_{am} is the intensity of the amorphous diffraction peak taken at 2θ between 18° and 19° for cellulose I.

3. Results and Discussion

3.1. Compositional Analysis

The raw materials (S and P) were analyzed to determine the cellulose content (Table 1).

Table 1. Composition of S and P from coconut fruit.

Composition	Shell (%w/w)	Pulp (%w/w)
Extractives	6.27	17.13
Cellulose	39.65	45.42
Hemicellulose	19.22	15.89
Lignin	30.80	17.90
Ash	4.06	3.66

A higher cellulose content (45.4%) is observed in the pulp (P), but the value obtained for the shell (S) is also important (39.6%). Comparing the lignocellulosic composition of P of the coconut fruit with the non-woody and woody raw materials commonly used to obtain cellulose, such as pinewood (37–43%), eucalyptus wood (41–50%), bamboo (43%), and bagasse (42–55%), similar and even higher cellulose contents are observed. These values open the expectations for the use of these lignocellulosic wastes from coconut fruits to obtain cellulose and, after a purification process, obtain MCC and MFC from them.

3.2. Cellulose Preparation and Purification

Table 2 shows the results of the yields after the alkaline and bleached treatments. The lower yield of alkaline treatment may be due to the presence of oil in both raw materials (S and P), mainly in the shell (S), as seen in the content of extractives in Table 1 regarding the bleaching treatment, the yield is quite promising for industrial scaling.

Table 2. Alkali and bleached treatment yields.

	Time (min)	Pulp (%w/w)	Shell (% w/w)
Alkaline treatment	240	20.52	30.88
Bleaching treatment	120	45.56	58.31

Figure 1 shows the DTG traces for S and P before and after alkaline and bleaching treatments.

The purification of cellulose was verified using DTG as the functional groups and characteristic peaks of hemicellulose (around 250 °C) and lignin (a small shoulder above 390 °C) are not seen, leaving only the characteristic cellulose peak (around 340 °C), intensified in both bleached samples. In addition, the effectiveness of alkali and bleaching treatments of the coconut residues was analyzed using FTIR spectroscopy by the decrease in or disappearance of peaks characteristics of hemicelluloses and lignin.

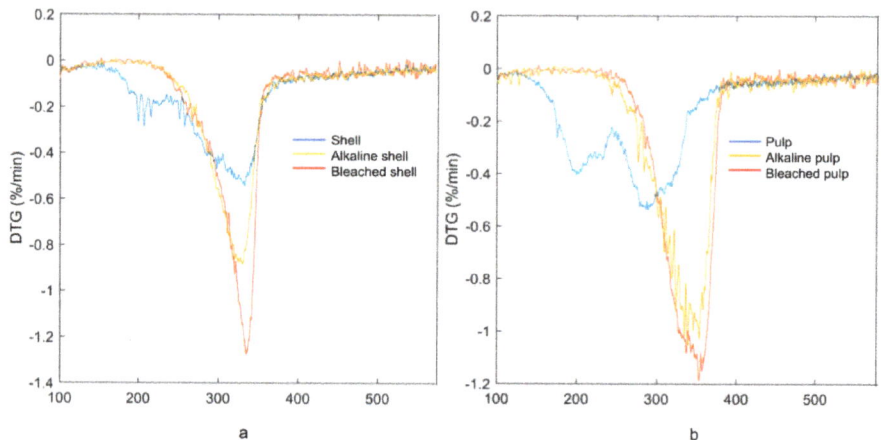

Figure 1. (**a**) DTG of S and (**b**) DTG of P before and after the alkaline and bleaching treatments.

3.3. Characterization of MCC and MFC

Fourier Transform Infrared Spectroscopy (FTIR)

Figure 2 shows the FTIR analyses performed on the MFC of the S at different times of mechanical treatment.

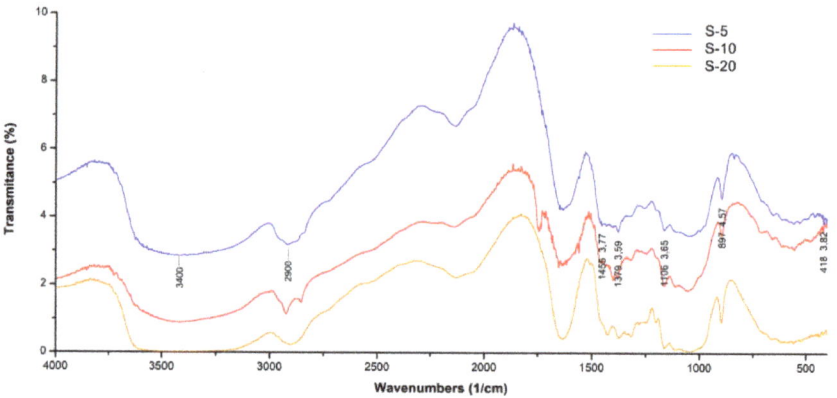

Figure 2. FTIR of the MFC of coconut shell (S) obtained at 5 (S-5), 10 (S-10), and 20 min (S-20).

Absorption bands close to 3400, 2900, 1430, 1370, 1160, and 897 cm^{-1} were associated with cellulose type I [17,18]. The peak at 2900 cm^{-1} was assigned to the CH stretching of the cellulose. The peak at 1640 cm^{-1} was due to the vibration of the adsorbed water molecules and also to the carbonyl groups, which may indicate the presence of hemicellulose [19,20]. The absorption peak at 1430 cm^{-1} was related to the crystalline band, which refers to the symmetric stretching of CH_2. The peak at 897 cm^{-1} was attributed to asymmetric stretching of the out-of-plane ring in cellulose due to β-glucoside bonding. The peak at 1640 cm^{-1} was associated with the H–O–H stretching vibration of carbohydrate-absorbed water. This indicates the presence of hemicellulose in MFCs, which favors fibrillation during mechanical treatment [20].

Other figures from the FTIR analysis for the MFC of the P and the MCC of the P and S show the same characteristic peaks of cellulose, indicating that purified MCs have been obtained using different treatments.

The crystallinity indices obtained using Equation (1) for the MCC are presented in Table 3.

Table 3. Crystallinity index for MCC and MFCs of coconut fruit.

	Sample	%CrI	Sample	%CrI
MCC	S15′	42.42	P15′	46.40
	S30′	68.19	P30′	64.36
	S60′	69.74	P60′	67.03

The peaks identified for the MCC samples correspond to Type I cellulose, and crystallinity index (%CrI) increased as the chemical treatment time increased, as observed in Table 3, for the S60 sample, with the 60 min of treatment. The increase in the crystalline index could be explained by the total or partial elimination of the hemicellulose and lignin structures, which reduced the amorphous region [21].

Regarding SEM analysis (Figure 3, a decrease in the diameter was verified for the MCFs as the mechanical treatment increased, being the best condition after 20 min treatment. The estimated diameter, after 20 min of treatment, was between 1 and 3 microns for the pulp (P) and 6 and 8 microns for the shell (S). In addition, a smaller size of the microfibers is evident in the MFCs obtained from the pulp compared to those obtained from the shell.

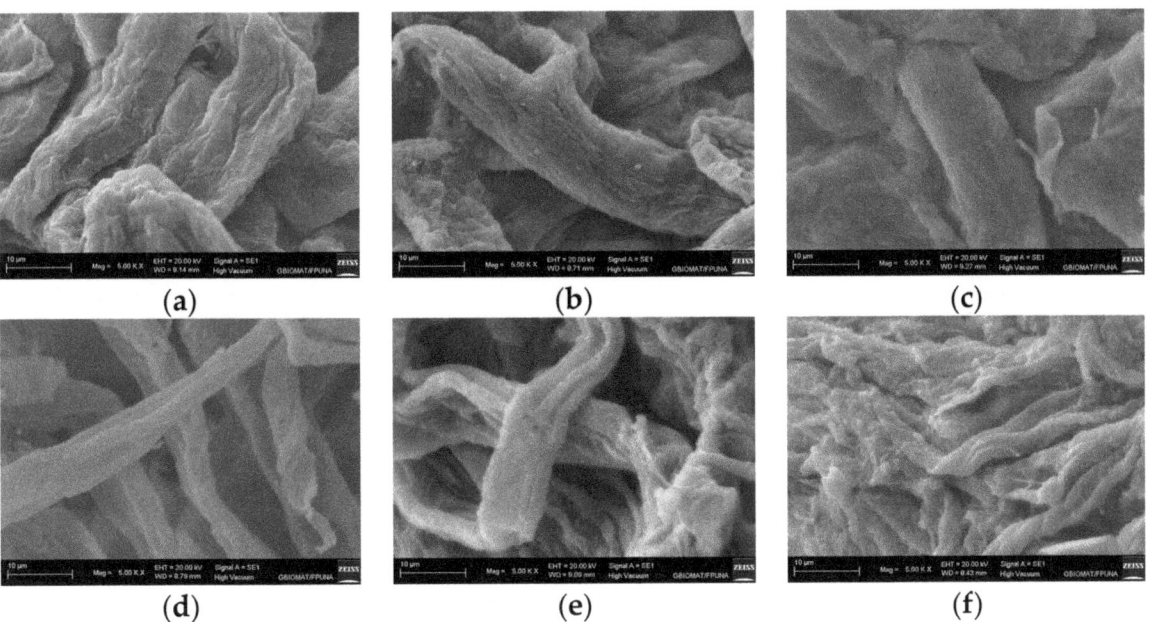

Figure 3. SEM micrographs of MFC from coconut fruit shell (a–c) and pulp (d–f) obtained after 5 min (a,d), 10 min (b,e), and 20 min (c,f) of mechanical treatment.

4. Conclusions

Coconut fruit wastes were valued by obtaining microcrystals (MCC) with high crystallinity index and cellulose microfibers (MFC) with high fibrillation. The purification treatments of the extracted cellulose were effective according to the characterizations carried out after the alkaline treatment and bleaching, verified via FTIR and TGA analysis.

In addition, high yields of purified cellulose for shell (S) and pulp (P) were obtained. Given their richness in cellulose and their (MCC) and (MFC) characteristics, S and P have the potential to be used as reinforcement for food packaging on the way to obtaining environmentally friendly materials.

Author Contributions: Conceptualization, S.D.; methodology, M.C.P., M.M. and S.D.; software, M.C.P.; validation, S.D. and M.M.; formal analysis, M.M. and S.D.; investigation, N.G. and P.A.C.; resources, S.D. and M.C.P.; data curation, M.C.P.; writing—original draft preparation, M.C.P. and S.D.; writing—review and editing, S.D. and M.M.; visualization, S.D.; supervision, S.D.; project administration S.D.; funding acquisition, S.D. All authors have read and agreed to the published version of the manuscript.

Funding: This research was funded by CONACYT under grant Code Project PINV18-514 and the Red ENVABIO100-CYTED (Ref. 121RT0108).

Institutional Review Board Statement: Not applicable.

Informed Consent Statement: Not applicable.

Data Availability Statement: Not applicable.

Acknowledgments: The authors thank Magna Monteiro (GBIOMAT, FP-UNA, Py) for assistance with SEM, FTIR, and XRD analysis. The authors gratefully acknowledge the Red Cyted ENVABIO100 121RT0108 for their financial support. S. Duarte would like to thank PRONII (CONACYT, PY).

Conflicts of Interest: The authors declare no conflict of interest.

References

1. Evaristo, A.B.; Grossi, J.A.S.; Carneiro, A.d.C.O.; Pimentel, L.D.; Motoike, S.Y.; Kuki, K.N. Actual and putative potentials of macauba palm as feedstock for solid biofuel production from residues. *Biomass-Bioenergy* **2016**, *85*, 18–24. [CrossRef]
2. Reis, S.B.; Mercadante-Simões, M.O.; Ribeiro, L.M. Pericarp development in the macaw palm Acrocomia aculeata (Arecaceae). *Rodriguésia* **2012**, *63*, 541–549. [CrossRef]
3. Plath, M.; Moser, C.; Bailis, R.; Brandt, P.; Hirsch, H.; Klein, A.-M.; Walmsley, D.; von Wehrden, H. A novel bioenergy feedstock in Latin America? Cultivation potential of Acrocomia aculeata under current and future climate conditions. *Biomass-Bioenergy* **2016**, *91*, 186–195. [CrossRef]
4. Ciconini, G.; Favaro, S.; Roscoe, R.; Miranda, C.; Tapeti, C.; Miyahira, M.; Bearari, L.; Galvani, F.; Borsato, A.; Colnago, L.; et al. Biometry and oil contents of Acrocomia aculeata fruits from the Cerrados and Pantanal biomes in Mato Grosso do Sul, Brazil. *Ind. Crop. Prod.* **2013**, *45*, 208–214. [CrossRef]
5. Ovelar, R.L.; Ortellado, J.; Echauri, C.; Aguero, J.; Galeano, M. Residuos de "acrocomia aculeata" como fuente de biomasa: Una revisión sistemática. *Extensionismo Innov. Y Transf. Tecnol.* **2019**, *5*, 326–330. [CrossRef]
6. Duarte, S.; Lv, P.; Almeida, G.; Rolón, J.C.; Perre, P. Alteration of physico-chemical characteristics of coconut endocarp—Acrocomia aculeata—By isothermal pyrolysis in the range 250–550 °C. *J. Anal. Appl. Pyrolysis* **2017**, *126*, 88–98. [CrossRef]
7. Oviedo Ch, A.; Vinueza, G.J. Lignocellulosic waste and its uses, a review. *infoANALÍTICA* **2020**, *8 (Extra 1)*, 133–147.
8. Hasan, M.; Lai, T.K.; Gopakumar, D.A.; Jawaid, M.; Owolabi, F.A.T.; Mistar, E.M.; Alfatah, T.; Noriman, N.Z.; Haafiz, M.K.M.; Khalil, H.P.S.A. Micro Crystalline Bamboo Cellulose Based Seaweed Biodegradable Composite Films for Sustainable Packaging Material. *J. Polym. Environ.* **2019**, *27*, 1602–1612. [CrossRef]
9. Huang, X.; Xie, F.; Xiong, X. Surface-modified microcrystalline cellulose for reinforcement of chitosan film. *Carbohydr. Polym.* **2018**, *201*, 367–373. [CrossRef] [PubMed]
10. Li, C.; Luo, J.; Qin, Z.; Chen, H.; Gao, Q.; Li, J. Mechanical and thermal properties of microcrystalline cellulose-reinforced soy protein isolate–gelatin eco-friendly films. *RSC Adv.* **2015**, *5*, 56518–56525. [CrossRef]
11. Wypych, G. Fillers—Origin, Chemical Composition, Properties, And Morphology. In *Handbook of Fillers*; ChemTec Publishing, Science Direct: Toronto, ON, Canada, 2016; pp. 13–266.
12. Ventura-Cruz, S.; Tecante, A. Nanocellulose and microcrystalline cellulose from agricultural waste: Review on isolation and application as reinforcement in polymeric matrices. *Food Hydrocoll.* **2021**, *118*, 106771. [CrossRef]
13. Álvarez, A.; Cachero, S.; González-Sánchez, C.; Montejo-Bernardo, J.; Pizarro, C.; Bueno, J.L. Novel method for holocellulose analysis of non-woody biomass wastes. *Carbohydr. Polym.* **2018**, *189*, 250–256. [CrossRef] [PubMed]
14. Browning, B.L. *Methods of Wood Chemistry Vol I & II*; John Wiley & Sons: New York, NY, USA, 1967.
15. Bhawna, S.; El Barbary, H.; Barakat, M. Chemical isolation and characterization of different cellulose nanofibers from cotton stalks. *Carbohydr. Polym.* **2015**, *134*, 581–589.
16. Moharram, M.A.; Mahmoud, O.M. FTIR spectroscopic study of the effect of microwave heating on the transformation of cellulose I into cellulose II during mercerization. *J. Appl. Polym. Sci.* **2008**, *107*, 30–36. [CrossRef]
17. Brahim, M.M.; El-Zawawy, W.K.; Jüttke, Y.; Koschella, A.; Heinze, T. Cellulose and microcrystalline cellulose from rice straw and banana plant waste: Preparation and characterization. *Cellulose* **2013**, *20*, 2403–2416. [CrossRef]
18. Razali, N.; Hossain, M.S.; Taiwo, O.A.; Ibrahim, M. Influence of Acid Hydrolysis Reaction Time on the Isolation of Cellulose Nanowhiskers from Oil Palm Empty Fruit Bunch Microcrystalline Cellulose. *BioResources* **2017**, *12*, 6773–6788. [CrossRef]
19. Adel, A.M.; El-Gendy, A.A.; Diab, M.A.; Abou-Zeid, R.E.; El-Zawawy, W.K.; Dufresne, A. Microfibrillated cellulose from agricultural residues. Part I: Papermaking application. *Ind. Crop. Prod.* **2016**, *93*, 161–174. [CrossRef]

20. Vora, R.; Shah, Y. Extraction, characterization of micro crystalline cellulose obtained from corn husk using different acid alkali treatment methods. *Indo Am. J. Pharm. Sci.* **2017**, *4*, 2399–2408.
21. Martinez-Pavetti, M.B.; Medina, L.; Espínola, M.; Monteiro, M. Study on two eco-friendly surface treatments on Luffa cylindrica for development of reinforcement and processing materials. *J. Mater. Res. Technol.* **2021**, *14*, 420–2427. [CrossRef]

Disclaimer/Publisher's Note: The statements, opinions and data contained in all publications are solely those of the individual author(s) and contributor(s) and not of MDPI and/or the editor(s). MDPI and/or the editor(s) disclaim responsibility for any injury to people or property resulting from any ideas, methods, instructions or products referred to in the content.

Proceeding Paper

Mechanical Properties of Pineapple Nanocellulose/Epoxy Resin Composites [†]

Gabriela Álvarez Véliz [1], Jorge Iván Cifuentes [1], Diego Batista [2], Mary Lopretti [3], Yendry Corrales [2], Melissa Camacho [2] and José Roberto Vega-Baudrit [2,4,*]

[1] Unidad de Investigación, Escuela de Ingeniería Mecánica, Universidad de San Carlos de Guatemala, Guatemala 01012, Guatemala; gaby.alvarez.usac@gmail.com (G.Á.V.); jicifuentes@ing.usac.edu.gt (J.I.C.)
[2] Laboratorio Nacional de Nanotecnología, Centro de Nacional de Alta Tecnología, LANOTEC-CENAT-CONARE, Pavas, San José 1615-1000, Costa Rica; dbatista@cenat.ac.cr (D.B.); ycorrales@cenat.ac.cr (Y.C.); kmce08@gmail.com (M.C.)
[3] Departamento de Técnicas Nucleares Aplicadas en Bioquímica y Biotecnología, CIN, Facultad de Ciencias, Universidad de la República, Montevideo 11400, Uruguay; mlopretti@gmail.com
[4] Laboratorio de Polímeros POLIUNA, Universidad Nacional, Heredia 86-3000, Costa Rica
* Correspondence: jvegab@gmail.com
[†] Presented at the 1st International Conference of the Red CYTED ENVABIO100 "Obtaining 100% Natural Biodegradable Films for the Food Industry", San Lorenzo, Paraguay, 14–16 November 2022.

Abstract: A study of materials for wind turbine blades with nanotechnology—from the energy point of view—is an essential topic because resources and fossil fuels are running out. Human beings need to create alternative energies, including wind energy. This research aims to improve the mechanical properties of epoxy resin wind turbine blades by incorporating nanocelluloses obtained from pineapple residues. To determine the quality of the nanobiocomposites, materials with different epoxy resin–nanocellulose ratios were prepared. The mechanical properties of tension, compression, and bending were evaluated, and hardness tests of the material were conducted. The results indicated a general improvement in all the mechanical properties considered over the material without the nanocellulose.

Keywords: blades; nanobiocomposites; nanocellulose; epoxy resin; pineapple waste

1. Introduction

The primary function of a wind turbine is to transform the kinetic energy into electrical energy produced by the movement of the blades of the wind turbine as a consequence of the passage of the wind through them. The wind circulates on both sides of the blades with different geometric profiles; a depression area is generated on the top face concerning the pressure on the whole face. Following this pressure, a resistance force is generated that opposes the movement, generating force in the rotor through kinetic energy. Turbine blades are elementary for the generation of electrical energy. They are in charge of receiving the power of the wind through speed [1,2].

The useful life and efficiency of the blades depend on their manufacturing. Ancient materials such as wood, steel, and aluminum have been used. Currently, they are manufactured with composite materials such as steel alloys, polyester, or epoxy resin reinforced with fiberglass or carbon fiber. The blades must be light in weight and have adequate mechanical behavior during their useful life. In this research, a material that comes from the agro-industrialization of pineapple cultivation is used. This fruit is widely distributed in Latin America and generates many problems for the environment [3–7].

Because of the extensive area coverage that pineapple cultivation represents, it is crucial to consider the large amount of waste or residues generated from the crop's production and industrial processing. The use and revalorization of these residues would avoid their inadequate disposal. This would be advantageous both from an economic point of view by reducing costs and from the perspective of the mitigation of environmental and health harms. There are methods for obtaining a material called nanocellulose, derived at the nanoscopic level from cellulose, which is part of the structure or plant cell wall of the pineapple and other plants. Cellulose is known to be the main component of the cell walls of plants; it is the most abundant biomaterial on the planet and provides remarkable properties of tensile strength [7–10].

This research aims to improve the mechanical properties of wind turbine blades. It is expected to obtain a composite material between the epoxy resin and the nanocellulose from pineapple peel waste. Incorporating this material aims to improve mechanical properties such as resistance, hardness, and modulus of elasticity, among others.

2. Materials and Methods

2.1. Materials

Pineapple wastes were supplied by Florida Products S.A., (Heredia, Costa Rica). Sodium hydroxide (NaOH), sodium hypochlorite (NaClO), clorhidric acid (HCl), sulfuric acid (H_2SO_4), and ethanol 95% reagents were obtained from Sigma-Aldrich (USA). A commercial epoxy resin (Hawk epoxy R1) with an epoxy resin catalyst (USA, Hawk epoxy C2) was used.

2.2. Nanocellulose Preparation

Pineapple peels were placed in a solution of 20 wt. % NaOH at 70–90 °C for one and a half hours, cleaned, and placed again in 12 wt. % NaOH for one hour. Next, they were bleached with a 2.5 wt. % NaClO solution at 60 °C for two hours. Afterward, white cellulose was treated with 17 wt. % HCl at 60 °C for two hours to obtain microcellulose. Finally, to obtain nanocrystalline cellulose, the acid hydrolysis of microcellulose was carried out using a solution of 65 wt. % H_2SO_4, a temperature of 55 °C, and constant stirring for 60 min. The samples were washed repeatedly with deionized water until they reached a neutral pH. Finally, nanocellulose suspension was dialyzed for 24 h to remove salt residues [6–8].

2.3. Composites Preparation

Epoxy resin with pineapple peel nanocellulose (0, 0.25, 0.5, 1, 2.5, 5, and 10 wt. %) was prepared. Epoxy resin catalyst was used. A glass mold of 10×10 cm was used to prepare samples. The mixtures were placed in the glass molds, separately.

Nanocellulose composites were prepared at 220 rpm for 8 min in a stirrer at room temperature. Samples were cured at 30 °C for 1 h, 60 °C for 30 min, and 100 °C for 2 h.

2.4. Sample Characterization

Samples were characterized using the Brinell hardness (HB) test (TH 1100 LEEB brand durometer), as well as tension (tensile strength), compression (compressive strength), bending (bending stress), and torsion (torsional strength) tests (Discovery DHR III Rheometer, TA Instruments). The samples for the analyses were prepared according to the ASTM protocols [11].

3. Results and Discussion

In general terms, it is observed that the addition of nanocellulose from pineapple peel waste generates positive results. The best results are observed by adding 0.25% up to 1% nanocellulose. At higher percentages, the mechanical properties decrease until reaching the brittle point of the material. Figure 1 shows the results of the Brinell hardness (HB) test. The Brinell scale characterizes the indentation hardness of materials through the scale of penetration of an indenter loaded on a material test piece. A pineapple nanocellulose content between 0.25 and 2.5% in the samples shows an increase in the BH parameter. Samples with 5% and 10% pineapple nanocellulose show a low BH value.

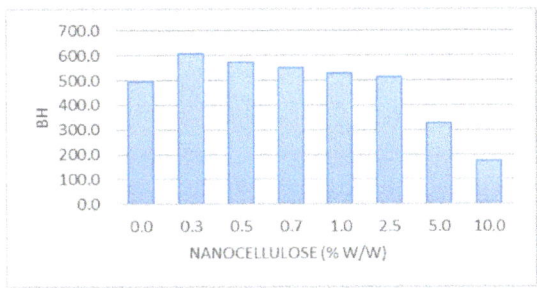

Figure 1. The Brinell hardness (HB) test in epoxy resin samples with nanocellulose obtained from pineapple wastes.

Similarly, Figure 2 shows the results of the tensile strength test in epoxy resin samples with nanocellulose obtained from pineapple wastes. The samples with a nanocellulose content between 0.25% and 1% showed higher tensile strength values than the nanomaterial. This indicates that the nanomaterial acts as a reinforcement for the epoxy matrix. In contrast, at higher values between 2.5% and 10%, it acts as a load that somewhat limits the material's mechanical properties.

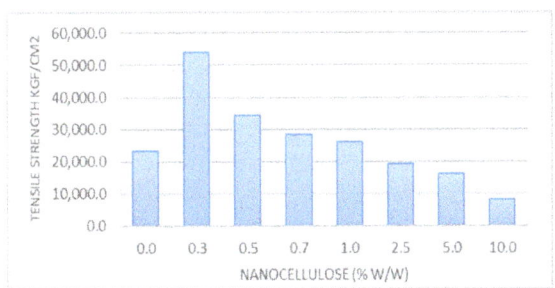

Figure 2. Tensile strength in epoxy resin samples with nanocellulose obtained from pineapple wastes.

Figure 3 shows the variation obtained from the compressive strength tests in the epoxy resin samples. In this case, this parameter significantly decreased for all samples with nanocellulose compared with samples without nanocellulose. The nanocellulose acts as a nanomaterial that decreases the internal pressure of epoxy resin. Samples with nanocellulose are compressed the easiest.

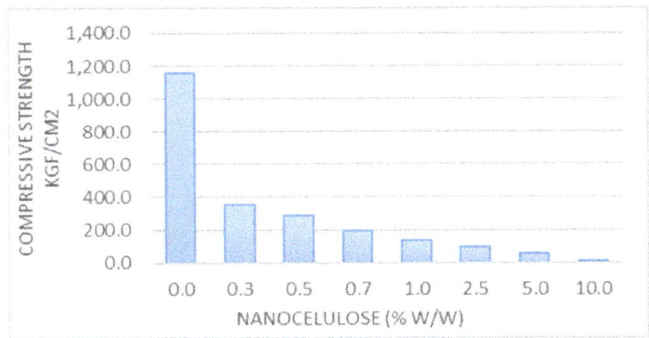

Figure 3. Compressive strength in epoxy resin samples with nanocellulose obtained from pineapple wastes.

Figure 4 shows the bending stress results obtained in all samples. Increased bending stress was observed in samples with 0.25 to 0.7% nanocellulose. Bending stress is a combination of compressive and tensile stresses. Thus, bending stress results are combined, as shown in Figures 2 and 3. Some samples showed an increase in bending stress, and some showed a dramatic decrease.

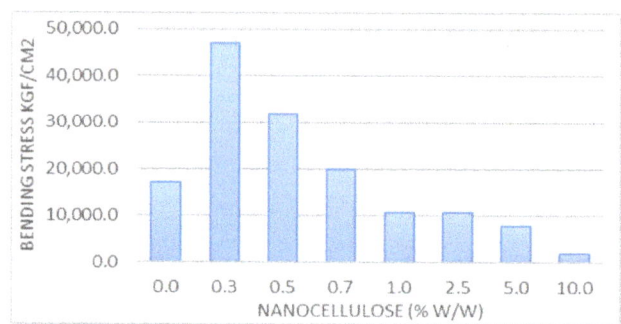

Figure 4. Bending stress test in epoxy resin samples with nanocellulose obtained from pineapple wastes.

Finally, in the torsional stress test (Figure 5), samples with a nanocellulose content between 0.25% and 0.7% show an increase in this property. In the same way, some samples showed a decrease. For example, the 5% and 10% nanocellulose samples showed minor torsional stress, as shown for bending stress.

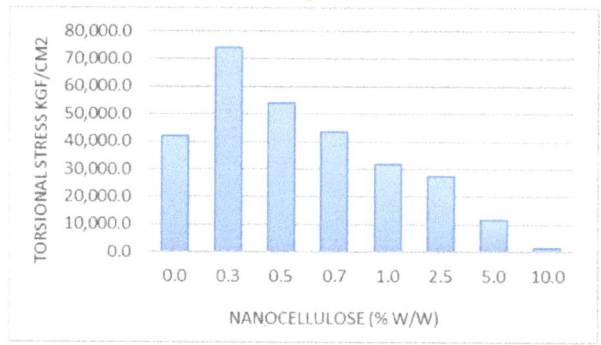

Figure 5. Torsional stress test in epoxy resin samples with nanocellulose obtained from pineapple wastes.

4. Conclusions

Positive results were found concerning the behavior of epoxy resin with inclusions of natural fiber as a nanocellulose base. The agricultural residues of pineapple, usually discarded or destroyed, proved to be a significant source of lignocellulosic biomass suitable for producing nanocellulose. As such, lignocellulosic materials are taken advantage of, and a double effect is achieved: the ecological benefit by eliminating a source of contamination and added economic value provided to the material.

The results indicated a general improvement in all the mechanical properties evaluated compared with the material without the pineapple nanocellulose. The mechanical properties in some epoxy resin samples were increased by incorporating nanocellulose in the polymer. This study could help improve the generation of alternative energies using plastic turbine blades reinforced with natural nanomaterials.

Author Contributions: Conceptualization, G.Á.V. and J.I.C.; methodology, J.I.C., D.B., M.C., M.L. and J.R.V.-B.; software, J.R.V.-B.; validation, M.C. and J.R.V.-B.; formal analysis, J.I.C.; investigation, G.Á.V., Y.C. and J.R.V.-B.; resources, J.R.V.-B.; data curation, J.I.C.; writing—original draft preparation, G.Á.V.; writing—review and editing, M.L. and J.R.V.-B.; visualization, J.I.C.; supervision, J.R.V.-B.; project administration, J.I.C.; funding acquisition, M.L. and J.R.V.-B. All authors have read and agreed to the published version of the manuscript.

Funding: This research received no external funding.

Institutional Review Board Statement: Not applicable.

Informed Consent Statement: Not applicable.

Data Availability Statement: Not applicable.

Acknowledgments: RED CYTED ENVABIO100 (Ref: 121RT0108) (interaction and publication cost).

Conflicts of Interest: The authors declare no conflict of interest.

References

1. Bronsted, P.; Lilhot, H.; Lystrup, A. Composite materials for wind power turbine blades. *Annu. Rev. Mater. Res.* **2005**, *35*, 505–538. [CrossRef]
2. Nørkær Sørensen, J. Aerodynamic Aspects of Wind Energy Conversion. *Annu. Rev. Mater. Res.* **2011**, *43*, 427–448.
3. Sánchez-Pardo, M.E.; Ramos-Cassellis, M.E.; Mora-Escobedo, R.; Jiménez-García, E. Chemical characterization of the industrial residues of the pineapple (*Ananas comosus*). *J. Agric. Chem. Environ.* **2014**, *3*, 53–56.
4. Morales-Vázquez, J.G.; López-Zamora, L.; Aguilar-Uscanga, M.G. Ethanol Production from Pineapple Waste. Ph.D. Thesis, Instituto Tecnológico de Orizaba, Orizaba, Mexico, 2020.
5. Segura, A.; Manriquez, A.; Santos, D.; Ambriz, E.; Casas, P.; Muñoz, A.S. Obtención de bioetanol a partir de residuos de cascara de piña (*Ananas comosus*). *Jovenes Cienc.* **2020**, *8*, 1–8.
6. Camacho, M.; Ureña, Y.R.C.; Lopretti, M.; Carballo, L.B.; Moreno, G.; Alfaro, B.; Baudrit, J.R.V. Synthesis and characterization of nanocrystalline cellulose derived from pineapple peel residues. *J. Renew. Mater.* **2017**, *5*, 271–279. [CrossRef]
7. Amores-Monge, V.; Goyanes, S.; Ribba, L.; Lopretti, M.; Sandoval-Barrantes, M.; Camacho, M.; Corrales-Ureña, Y.; Vega-Baudrit, J.R. Pineapple Agro-Industrial Biomass to Produce Biomedical Applications in a Circular Economy Context in Costa Rica. *Polymers* **2022**, *14*, 4864. [CrossRef] [PubMed]
8. Rambabu, N.; PAnthapulakkal, S.; Sain, M.; Dalai, A.K. Production of nanocellulose fibers from pinecone biomass: Evaluation and optimization of chemical and mechanical treatment conditions on mechanical properties of nanocellulose films. *Ind. Crops Prod.* **2016**, *83*, 746–754. [CrossRef]
9. Mahardika, M.; Abral, H.; Kasim, A.; Arief, S.; Asrofi, M. Production of nanocellulose from pineapple leaf fibers via high-shear homogenization and ultrasonication. *Fibers* **2018**, *6*, 28. [CrossRef]
10. Moon, R.J.; Schueneman, G.T.; Simonsen, J. Overview of cellulose nanomaterials, their capabilities and applications. *JOM* **2016**, *68*, 2383–2394. [CrossRef]
11. *ASTM D638-14*; Standard Test Method for Tensile Properties of Plastics. ASTM International: West Conshohocken, PA, USA, 2022. [CrossRef]

Disclaimer/Publisher's Note: The statements, opinions and data contained in all publications are solely those of the individual author(s) and contributor(s) and not of MDPI and/or the editor(s). MDPI and/or the editor(s) disclaim responsibility for any injury to people or property resulting from any ideas, methods, instructions or products referred to in the content.

Proceeding Paper

Biofilms Functionalized Based on Bioactives and Nanoparticles with Fungistatic and Bacteriostatic Properties for Food Packing Uses [†]

Gabriela Lluberas [1,*], Diego Batista-Menezes [2], Juan Miguel Zuñiga-Umaña [2], Gabriela Montes de Oca-Vásquez [2,3], Nicole Lecot [1], José Roberto Vega-Baudrit [2] and Mary Lopretti [1,*]

[1] Laboratory of Nuclear Techniques Applied to Biochemistry and Biotechnology, Nuclear Research Center, Faculty of Sciences, Universidad de la República, Montevideo 11400, Uruguay; nlecot@fcien.edu.uy

[2] National Nanotechnology Laboratory, National Center for High Technology, Pavas, San José 10109, Costa Rica; dbatista@cenat.ac.cr (D.B.-M.); jzuniga@cenat.ac.cr (J.M.Z.-U.); mmontesdeoca@utn.ac.cr (G.M.d.O.-V.); jvegab@gmail.com (J.R.V.-B.)

[3] Center for Sustainable Development Studies, Universidad Técnica Nacional, Alajuela 1902-4050, Costa Rica

* Correspondence: gabriela_lluberas@hotmail.com (G.L.); mlopretti@gmail.com (M.L.)

[†] Presented at the 1st International Conference of the Red CYTED ENVABIO100 "Obtaining 100% Natural Biodegradable Films for the Food Industry", San Lorenzo, Paraguay, 14–16 November 2022.

Abstract: The objective of this work was to formulate PVA films with the addition of modified phenols, chitosan, silver, or copper nanoparticles with fungistatic and bacteriostatic activity. The films were characterized by Fourier transform infrared spectroscopy, thermogravimetry analysis, differential scanning calorimetry, and scanning electron microscopy. The bacteriostatic activity against *Salmonella gallinarum* and *Lactobacillus acidophilus*, and the fungistatic activity against *Penicillium acidophilus* were determined. The results indicated that the addition of phenols enhanced the effect on the stability of the chemical structure of the PVA film. PVA films with modified bioactives and nanoparticles inhibited the colonization of the microorganisms tested, indicating germicidal control.

Keywords: food packaging films; chitosan; PVA; nanoparticles; phenols extract

1. Introduction

Worldwide, numerous efforts are being made to reduce the use and environmental impact of petroleum-derived plastics, which has led to an increase in research aimed at obtaining biodegradable materials with special functionalities that allow their use in different sectors, such as environmental, food, medicine, and agriculture, among others [1].

The raw materials derived from renewable resources are biodegradable. The materials used for the formation of biodegradable films are mainly composed of cellulose, chitosan, starch, dextrins, alginates, and pectins [2–5]. In general, films made from biopolymers are sensitive to environmental conditions, especially relative humidity, and have low mechanical strength even when protein films have high elasticity [6]. A possible alternative to improve the mechanical characteristics of protein-based films could be the mixture of these biopolymers with synthetic polymers, such as polyvinyl alcohol (PVA), which is hydrophilic and biodegradable [7]. Some studies on the development and characterization of films based on mixtures of PVA and proteins, such as PVA/wheat gluten [8], PVA/hydrolyzed collagen [9], and PVA/gelatin [10], among others.

In the area of food packaging, it is very important to preserve the food quality and nutritional value for consumer safety [11]. Recently, it has been proposed to incorporate different nanomaterials and bioactive compounds into edible films to improve the film's physico-chemical, mechanical, and thermal characteristics and provide other properties such as antimicrobial films without causing environmental problems [12,13].

Metallic nanoparticles (MNPs) constitute a special and particularly valuable group of nanoparticles (NPs) with interesting properties in the protection, preservation of food, and extension of the shelf life of food [14–17]. Other bioactives like polyphenols from lignins and tannins constitute a special valuable group of micro- and nanoparticles (NPs) with interesting properties in the protection and preservation of food, among other applications [1,15,16,18–21].

Chitosan (CHT) is recognized for its effective germicidal action, with a broad microbial spectrum and low toxicity in animals and humans [22,23]. The polycationic structure of chitosan plays a fundamental role in electrostatic fixation and bacterial deactivation. The bactericidal activity of functionalized chitosan films combined with polyvinyl alcohol has been reported [7]. Regarding the capacity of antifungal activity, it suppresses sporulation and germination, inhibiting its growth [7,23].

The objective of this work is to formulate different films based on PVA functionalized with different bioactives (CHT and phenols), AgNPs, and CuNPs. The morphological, physical-chemical, and thermal properties and germicidal activity of the films were evaluated. The germicidal properties in microorganisms were evaluated, which are found in greater proportion in the daily environment and in some pathogens that affect food without adequate conservation.

2. Materials and Methods

2.1. Materials

The materials used were glutaraldehyde and spam 80 from drogueria Paysandu (Uruguay) and polyvinyl alcohol (PVA) from Acros Organics (USA). The chitosan 1% p/v, AgNPs, CuNPs, and phenols were supplied from the Nuclear Research Center, Faculty of Science, University of the Republic.

2.2. Preparation of PVA Films

First, 10 mL of 10% PVA solution was prepared and dissolved in water, and then 1 mL of Spam 80 was added with vigorous agitation. Subsequently, 8 mL of glutaraldehyde was slowly added (dropwise) with continuous agitation. The solutions were cast on a Petri dish and maintained under constant agitation for 48 h at room temperature until solidification. Then, 0.5 mL of AgNPs (1 mM and 2 mM), CuNPs (1 mg/mL), chitosan NPs (1 mg/mL), and phenol NPs (2 mg/mL) were added in order to increase the germicidal properties of the PVA films. The codes and descriptions are given in Table 1.

Table 1. Film sample codes and descriptions.

Codes	Description
PVA	PVA film = control
PVA-PH	PVA film with 0.05 mg/mL of phenols
PVA-CH-PH	PVA film with 0.025 mg/mL of chitosan and 0.05 mg/mL of phenols
PVA-AgNPs	PVA film with 0.025 mg/mL and 0.05 mg/mL of AgNPs
PVA-CuNPs	PVA film with 0.025 mg/mL of CuNPs

2.3. Films Characterization

2.3.1. Scanning Electron Microscopy

The films were analyzed using a scanning electron microscope (JSM-6390LV, Jeol, LANOTEC, San Jose, Costa Rica), with a voltage acceleration of 10 kV, secondary electrons (SEI), and a spot size of 40. Images were taken at different magnifications to identify the morphology of the films.

2.3.2. Fourier-Transform Infrared Spectroscopy

FTIR spectra of the films were recorded using an FTIR Nicolet 6700 spectrophotometer with a diamond ATR module (Thermo Fisher Scientific, Miami, FL, USA) in the number range waveform from 4000 to 500 cm^{-1} with a standard resolution of 4 cm^{-1} and a scanning speed of 32 cm^{-1}/s. The results were analyzed using the OMNIC 8.1 software (OMNIC Series 8.1.10, Thermo Fisher Scientific).

2.3.3. Thermogravimetric Analysis

The films were analyzed via thermogravimetric analysis (TGA) TGA-Q500 (TA Instruments, Philadelphia, PA, USA) equipped with Universal Analysis 2000 software (version 4.5A, TA Instruments, USA). Approximately 5 mg of the sample was used for the TGA analyses, with a temperature ramp of 10 °C/min from 25 to 800 °C under nitrogen (flow rate 90 mL/min).

2.3.4. Differential Scanning Calorimetry

Differential scanning calorimetry (DSC) was performed using the DSC Q200 equipment, TA Instruments (USA). The samples were analyzed in duplicate using a ramp that covered a cycle from 25 °C to 400 °C, with a heating rate of 10 °C/min and a nitrogen flow of 10 mL/min. The data obtained were analyzed using the TA Universal Analysis software (Advantage Software v5.5.24).

2.4. Determination of Control of Environmental Microorganisms

2.4.1. Determination of Bacteriostatic Activity against *Lactobacillus*

For the test, pure cultures of *Lactobacillus acidophilus* were maintained in 10 mL of liquid medium with 20% w/v glucose and 7% w/v yeast extract in a 50 mL flask. The pH was adjusted to 5.5. Then, it was incubated at 30 °C for 48 h without shaking. Subsequently, 0.1 mL of the inocula was plated on Petri dishes with solid culture medium MRS and films (PVA, PVA-PH, PVA-CH-PH, PVA-AgNPs, and PVA-CH-CuNPs) of 2 × 2 cm were incubated in an oven at 33 °C. The bacterial growth was monitored at 24 and 48 h ($n = 2$).

2.4.2. Determination of Bacteriostatic Activity against *Salmonella gallinarum*

A reconstituted *Salmonella gallinarum* strain was used for this determination. The bacteria were cultured at 37 °C for 24 h. Then, 10 µL of the inocula was plated in Petri dishes containing TSA culture medium. Subsequently, seeding was spread from one end to the middle of the plate surface. Then, a piece of membrane 2 × 2 cm in diameter (PVA, PVA-PH, PVA-CH-PH, PVA-AgNPs, and PVA-CH-CuNPs) was placed in the center of the plate and incubated in an oven at 37 °C. The bacterial growth was monitored until the 7th day post-incubation.

2.4.3. Determination of Fungistatic Activity against *Penicillium*

For the test, pure cultures of *Penicillium* sp. were maintained in 10 mL of liquid medium with 20% w/v glucose and 7% w/v yeast extract in a 50 mL flask. The pH was adjusted to 6.0. Then, it was incubated at 37 °C for 7 days under shaking. Subsequently, 0.1 mL of the inocula was plated on Petri dishes containing solid culture medium MRS and films (PVA, PVA-PH, PVA-CH-PH, PVA-AgNPs, and PVA-CH-CuNPs) of 2 × 2 cm and incubated at 33 °C. The fungal growth was monitored on days 5, 7, and 10 ($n = 2$). Some photographs were taken to illustrate the effect.

3. Results and Discussion

3.1. Films Characterization

The morphology of the films (PVA, PVA-PH, PVA-CH-PH, PVA-AgNPs, and PVA-CuNPs) were analyzed using SEM (Figure 1). The morphology of the surface and cross-section of the PVA films was smooth with some pores between 5 and 20 µm and a thickness of 110 µm. The PVA-PH film had a rough surface with pores between 20 and 100 µm and a thickness of 200 µm. Otherwise, the PVA-CH-PH film had a smooth surface with pores between 1 and 25 µm and a thickness of 100 µm. The PVA-AgNPs film presented a rough surface with microparticles in the presence of pores with sizes between 1 and 15 µm; this film had a thickness of 150 µm. Similar to the PVA-AgNPs, the PVA-CuNPs film presented a rough surface with microparticles between 5 and 20 µm on the surface, and the thickness of this film was 50 µm.

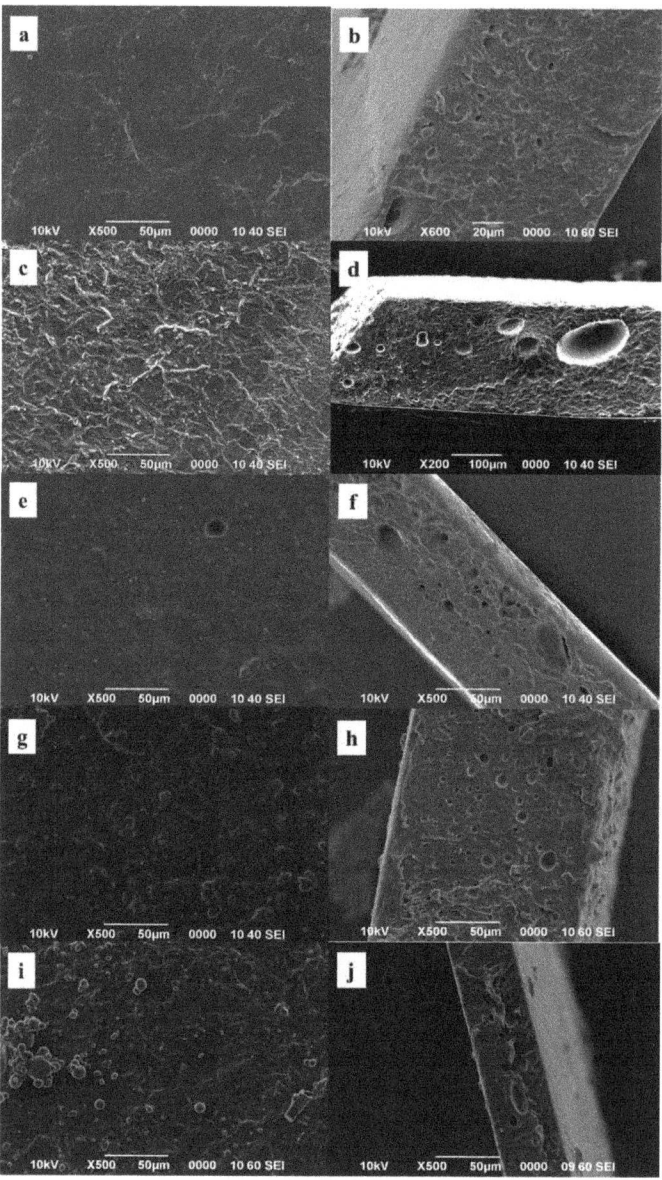

Figure 1. SEM surface (**a**,**c**,**e**,**g**,**i**) and cross-sectional (**b**,**d**,**f**,**h**,**j**) images of (**a**,**b**) PVA, (**c**,**d**) PVA-PH, (**e**,**f**) PVA-CH-PH, (**g**,**h**) PVA-AgNPs, and (**i**,**j**) PVA-CuNPs.

3.1.1. Fourier Transform Infrared Spectroscopy Analysis

FTIR-ATR analysis was carried out to study the molecular interaction between CH, PVA, AgNPs, CuNPs, and natural phenols in the films.

Figure 2 shows the FTIR spectrum of the PVA-based polymeric film samples, where it was possible to observe a broad absorption band at 3014–3680 cm^{-1} that was attributed to the stretching vibration of OH groups, which may be due to the presence of hydroxyl groups and residual moisture content of PVA films [24,25]. On the other hand, the nearby bands at 2920 and 2854 cm^{-1} were attributed to the asymmetric and symmetric stretching of the CH and CH$_2$ groups, respectively. In addition, a peak was observed at 1740 cm^{-1}, which was due to the stretching of the C-O group [3].

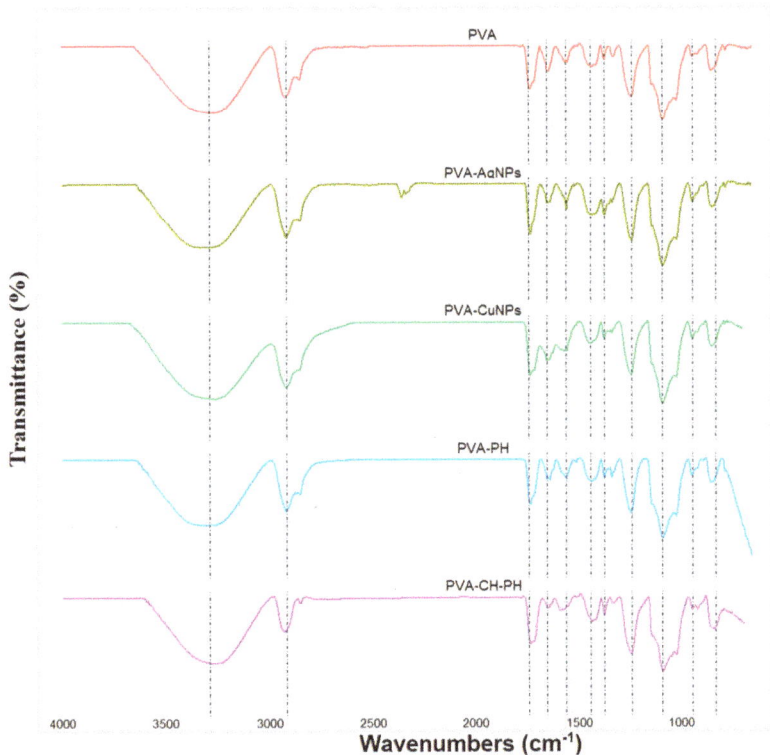

Figure 2. FTIR spectra of PVA, PVA-AgNPs, PVA-CH-PH, PVA-CuNPs, and PVA-PH films.

The peaks at 1563 cm^{-1} are designated as the stretching vibrations of the C=O group, while the peaks near 1420, 1370/1330/1242, 1086, and 838 cm^{-1} are attributed to the in-plane bending of the OH, the bending of the CH group, and the stretching of the C-O and CH of the PVA, respectively [11,25–28]. In addition, an absorption peak was observed at 1142 cm^{-1} in the long band between 1085 and 1150. According to the literature, this vibrational band is mainly attributed to the PVA crystallinity, which is related to the carboxyl (C-O) stretch band [24,29,30]

According to the results, it was clearly observed that the main peaks of the spectra are mainly associated with polyvinyl alcohol. This is due to the high concentration of this material, which is the polymeric matrix of the films, depending on the other components present. However, the bands at 1420–1375, 1085–1150, and 838 cm^{-1}, which are also characteristic of chitosan, may be interposed with those of PVA. In this case, for chitosan, these bands are attributed to the symmetric deformations of the CH2 and CH3 groups, the stretching vibrations of the C–O–C groups characteristic of polysaccharides, and the amine groups of chitosan, respectively [31–33].

3.1.2. Thermogravimetric Analysis

Figure 3 shows the TGA and derivative of TGA (DTG) of PVA-PH, PVA-CH-PH, PVA-AgNPs, and PVA-CuNPs films. All the films show similar curves with four steps of weight loss as a function of the increasing temperature. The first step between 25 and 125 °C corresponds to the evaporation of water [34]. The second step was the main degradation region at 150–350 °C due to the degradation of the exposed side chain and the breakdown of polymeric chains of PVA and chitosan [35]. The third weight loss (350–425 °C) corresponds to the oxidative decomposition of carbon residues, [35] and the final weight loss (425–500 °C) represents the decomposition to ash [27,34,36].

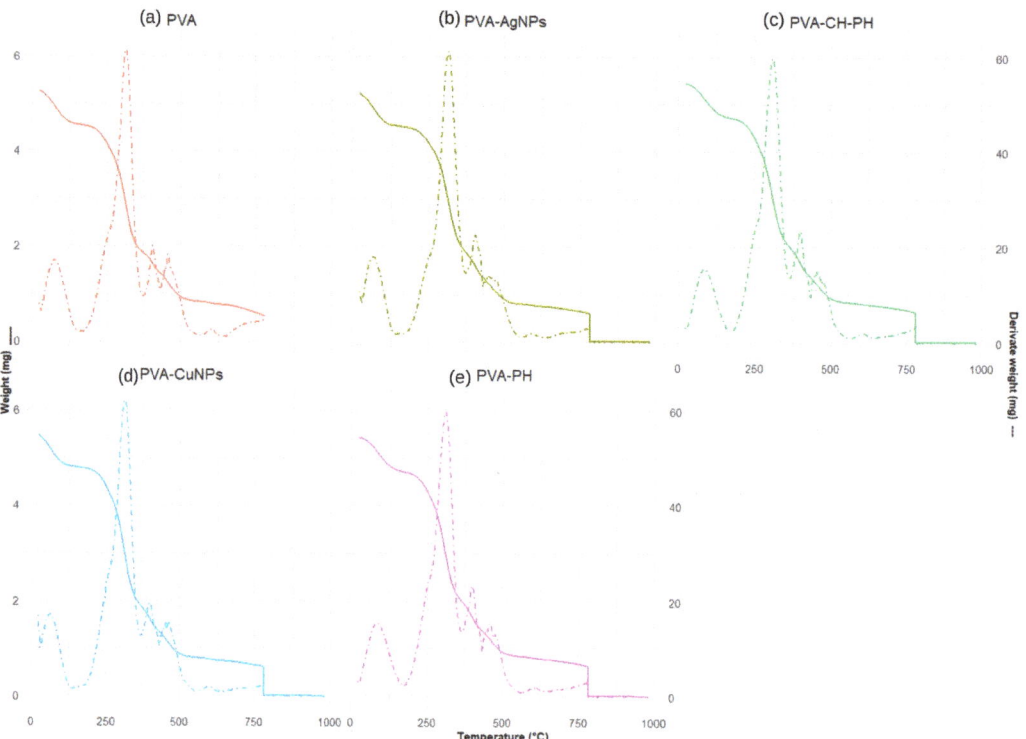

Figure 3. TGA and DTG curves of (**a**) PVA, (**b**) PVA-AgNPs, (**c**) PVA-CH-PH, (**d**) PVA-CuNPs, and (**e**) PVA-PH films.

According to the thermal profiles of the samples, there are no significant differences in the thermogram curves. This indicates that the different additives applied in the formulations do not improve the thermal stability parameters. This is possibly associated with the low concentration of the additives and/or the low degradation temperature, which may be involved in the main PVA decomposition events.

3.1.3. Differential Scanning Calorimetry

Figure 4 shows the DSC curves of the PVA, PVA-AgNPs, PVA-CH-PH, PVA-CuNPs, and PVA-PH films as a function of temperature. The results showed endothermic events in different temperature ranges. The first is reported near 45 °C and corresponds to the glass transition temperature (Tg) of the PVA polymer [37]. On the other hand, events between 60 and 170 °C represent dehydration processes due to the evaporation of physically adsorbed water content [35]. Finally, the events that appear near 175–195 °C and 240–350 °C can be attributed to the melting point (Tm) and thermal pyrolysis of the PVA polymer, respectively, which agrees with what has been reported by other authors [27,38,39].

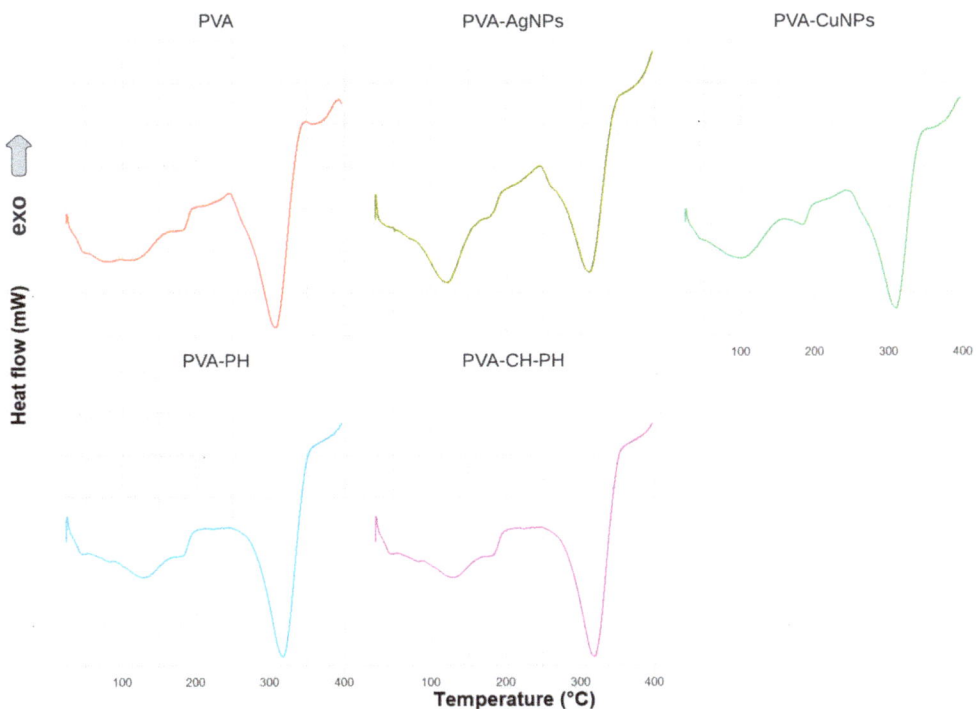

Figure 4. DSC curves of PVA, PVA-AgNPs, PVA-CuNPs, PVA-PH, and PVA-CH-PH films.

According to the results, it was possible to observe a displacement of +10 °C in the thermal pyrolysis events for the PVA-PH film compared to the PVA film. This could be associated with the molecular interactions between PVA and phenols, which in turn have a more stable chemical structure.

3.2. Fungistatic and Bacteriostatic Activity

The results of bacteriostatic activity of the films against *Lactobacillus* showed that the PVA film presented moderate bacterial growth on the films; in the PVA-AgNPs films at a concentration of 1 mM and 2 mM, the bacterial growth was minimal and null, respectively. Similarly, in the PVA-CuNPs and PVA-PH, the growth was minimal at 2% and 1% of the plate, respectively. The highest inhibition activity was observed for the PVA-CH-PH film, where there was a synergy between the germicidal activity of the two compounds. It should also be mentioned that there was no invasion of the microorganism on the film in any of the cases, and there was actually an inhibition of growth (Figure 5).

We observed the growth inhibition of *Salmonella* in PVA-AgNPs 2 mM films, whereas, in PVA-AgNPs 1 mM films, total bacterial growth on the film was observed. The antibacterial activity of the nanocomposite films was also investigated against *Salmonella typhimurium* using the disk diffusion method. The results showed that the PVA film has excellent antibacterial activity against *Salmonella typhimurium*, as reported in other works [40,41]. In the PVA-PH and PVA-CH-PH films, we determined the growth inhibition of *Salmonella* sp., whereas, in the case of the PVA-CuNPs film, the growth was over the entire edge of the film (Figure 6).

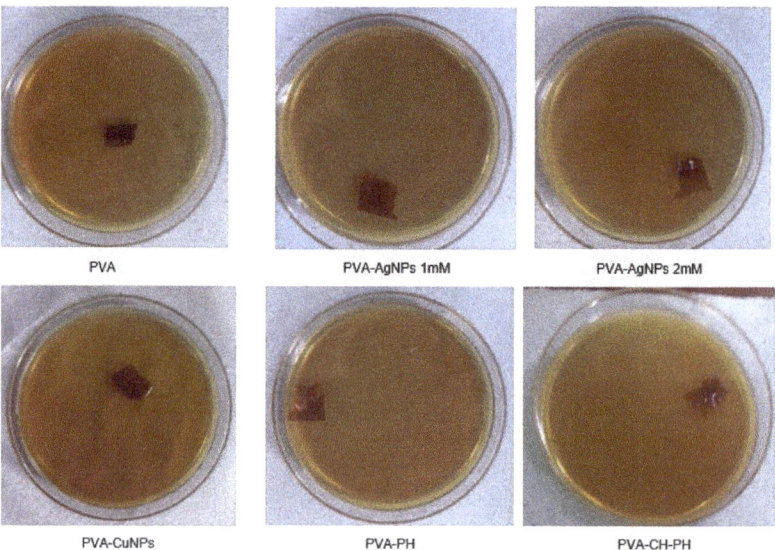

Figure 5. Culture and inhibition control of *Lactobacillus* sp. after 7 days of contact with PVA; PVA-AgNP 1 mM; A-AgNP 2 mM; PVA-CuNPs; PVA-PH; and PVA-CH-PH films.

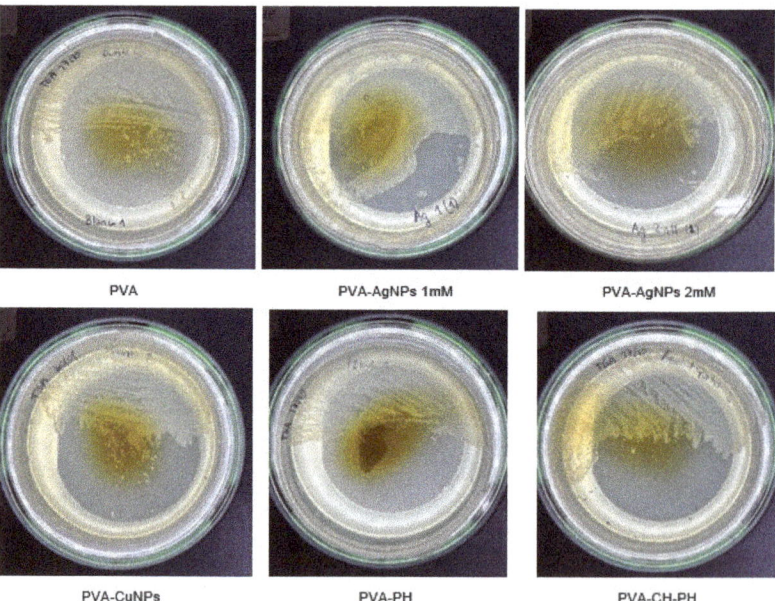

Figure 6. Culture and inhibition control of *Salmonella* after 7 days of contact with PVA; PVA-AgNPs 1 mM; PVA-AgNPs 2 mM; PVA-CuNPs; PVA-PH; PVA-CH-PH films.

Other authors determined the antibacterial and antifungal activities of nanocomposite PVA films [40] and PVA-Starch film with the addition of oregano essential oil [42]. Moreover, the microbial activity of the films with the addition of AgNPs was determined in films with the addition of these NPs [40,43]. In addition, chitosan was used due to its germicidal activity in PVA films [40].

The results of the fungistatic activity against *Penicillium* sp. (Figure 7) showed that in all the treatments, there was no fungal growth on the culture media or film after 7 days of incubation. The cultures were kept under study for a longer time to test resistance and control over time. The behavior at 21 days showed a slight minimum growth in the PVA film, whereas in the others, the growth did not manifest.

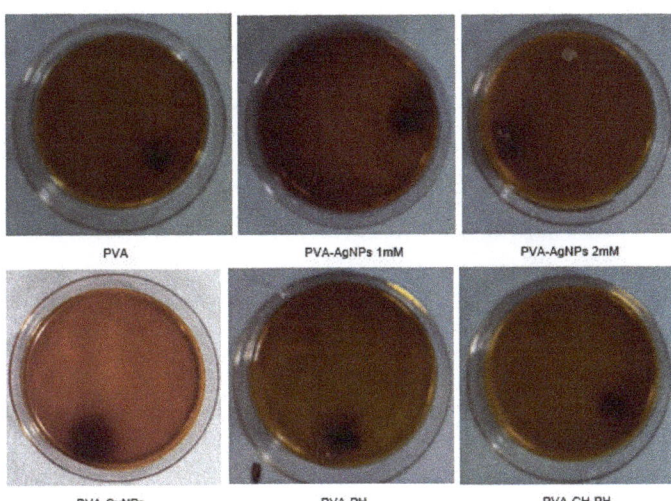

Figure 7. Culture and inhibition control of *Penicillium* after 7 days of contact with PVA; PVA-AgNP 1 mM; PVA-AgNP 2 mM; PVA-CuNPs; PVA-PH; and PVA-CH-PH films.

The fungistatic and bacteriostatic activity of the phenols incorporated in the PVA-PH films developed in this research had similar behavior to that reported by other authors [44], who found antimicrobial activity in films using bioactives such as phenols of plant extracts and propolis, without being incorporated into films. This form of direct application of bioactives produces diffusion, which produces inhibitory activity by halo production.

It has been determined in the literature that phenolic compounds such as phenol, hexachlorophenol, and thymol present an intermediate level of disinfection, and their activity is closely related to the concentration and the microbial species to be treated. Their mechanism of action is via the disruption of the cell wall and membrane and inactivation of enzyme systems. Moreover, Abud-Blanco et al. [19] determined that films with sulfonated phenolic compounds showed antibacterial activity against *Enterococcus feacalis*. Otherwise, tannins obtained from plant extracts were reported as antimicrobial agents for both Gram-positive bacteria such as *Listeria monocytogenes* and *Staphylococcus aureus*, and Gram-negative bacteria *Escherichia coli* and *Salmonella enterica* Serovar Typhimurium [45].

Other authors [18] described the germicidal properties of propolis and its phenolic components, like caffeic acid, ferulic acid, isoferulic acid, 3,4-dimethoxycinnamic acid, pinobanksin, caffeic acid benzyl ester, and caffeic acid phenethyl ester, which present antimicrobial properties and inhibitory effect on different bacteria such as *E. coli*, *Lactobacillus plantarum*, *S. aureus*, *S. epidermidis*, *Pseudomonas aeruginosa*, *P. fluorescens*, *Listeria monocytogenes*, *L. innocua*, *Klebsiella pneumoniae*, *Salmonella typhimurium*, *S. enteritidis*, *Streptococcus agalactiae*, *S. mutans*, *Bacillus cereus*, *B. subtilis*, *Citrobacter freundii*, *Enterobacter aerogenes*, *Shigella dysenteriae*, *Yersinia enterocolitica*, and *Pantoea agglomerans*.

The antimicrobial tests carried out on the PVA-CuNPs film showed an inhibition since the bacterial strains did not grow on the film. They reach the edge in some cases or present an inhibition halo of about 2 mm. Other studies in which the NPs were used directly were reported by the authors of [17], who observed inhibition halos of the positive control for Staphylococcus epidermidis, Aerococcus viridans, Ochrobactrum anthropi, and

Micrococcus lylae with diameters of inhibition of 21, 18, 7, and 19 mm, respectively, whereas for the negative control, the diameters of inhibition were 6 mm for all the bacteria studied. Many researchers affirm that the essential mechanism for the cytotoxicity generated by NPs is the release of Cu^{2+} ions that react with the thiol (SH) groups of the proteins present on the surface of the bacterial cell membrane. These proteins protrude from the cell membrane, allowing the transport of nutrients through the cell wall. The NPs can inactivate these proteins, thereby reducing the membrane permeability *and* causing cell death [17].

It is important to emphasize that none of the developed films grew microorganisms on the film, which indicates that the bioactives and NPs added to the PVA films had great germicidal control over the microorganisms studied. Future work on the study of other genera of microorganisms that are present in food or food packaging can be conducted, and it is also important to study the permanence of bioactives and NPs in the film since it would be desirable that they not be released in the food to be preserved.

4. Conclusions

The films obtained can be of different thicknesses, thin for coating, or of greater caliber for container food packaging. The methodology for obtaining films was adequate to obtain PVA membranes that are physically and chemically similar to each other and have a low production cost. The incorporation of phenolic, NPs, and chitosan bioactives into PVA showed a synergistic effect on antibacterial and fungicidal activity. Based on this study, the modified films that incorporate different bioactives can be applied as novel potential food packaging materials.

Author Contributions: M.L.: Conceptualization, methodology, investigation, resources, writing—original draft preparation, writing—review and editing, supervision, project administration, and funding acquisition; J.R.V.-B.: visualization, supervision, project administration, and funding acquisition; D.B.-M.: methodology, software, validation, writing—original draft preparation, writing—review and editing, visualization, and supervision; G.M.d.O.-V.: formal analysis, writing—original draft preparation, writing—review and editing, visualization, and supervision; J.M.Z.-U.: methodology, software, validation, formal analysis, writing—original draft preparation, and writing—review and editing. N.L.: formal analysis, visualization, writing—review and editing, and investigation. G.L.: methodology, investigation, software, validation, writing—original draft preparation, writing—review and editing, visualization, and supervision. All authors have read and agreed to the published version of the manuscript.

Funding: This research was funded by ENVABIO100 (Ref: 121RT0108).

Institutional Review Board Statement: Not applicable for studies not involving humans or animals.

Informed Consent Statement: Not applicable for studies not involving humans.

Data Availability Statement: Data sharing is not applicable to this article.

Conflicts of Interest: The authors declare no conflict of interest.

References

1. Lopretti, M.; Lecot, N.; Rodriguez, A.; Lluberas, G.; Orozco, F.; Bolaños, L.; Montes De Oca, G.; Cerecetto, H.; Vega-Baudrit, J. Biorefinery of rice husk to obtain functionalized bioactive compounds. *J. Renew. Mater.* **2019**, *7*, 313–324. [CrossRef]
2. Shiekh, K.A.; Liangpanth, M.; Luesuwan, S.; Kraisitthisirintr, R.; Ngiwngam, K.; Rawdkuen, S.; Rachtanapun, P.; Karbowiak, T.; Tongdeesoontorn, W. Preparation and characterization of bioactive chitosan film loaded with cashew (*Anacardium occidentale*) leaf extract. *Polymers* **2022**, *14*, 540. [CrossRef] [PubMed]
3. Kochkina, N.E.; Lukin, N.D. Structure and properties of biodegradable maize starch/chitosan composite films as affected by PVA additions. *Int. J. Biol. Macromol.* **2020**, *157*, 377–384. [CrossRef]
4. Rafique, A.; Mahmood Zia, K.; Zuber, M.; Tabasum, S.; Rehman, S. Chitosan functionalized poly(Vinyl alcohol) for prospects biomedical and industrial applications: A review. *Int. J. Biol. Macromol.* **2016**, *87*, 141–154. [CrossRef] [PubMed]
5. Tharanathan, R.N. Biodegradable films and composite coatings: Past, present and future. *Trends Food Sci. Technol.* **2003**, *14*, 71–78. [CrossRef]
6. Chen, H.; Wang, J.; Cheng, Y.; Wang, C.; Liu, H.; Bian, H.; Pan, Y.; Sun, J.; Han, W. Application of protein-based films and coatings for food packaging: A review. *Polymers* **2019**, *11*, 2039. [CrossRef]

7. Annu; Ali, A.; Ahmed, S. Eco-friendly natural extract loaded antioxidative chitosan/polyvinyl alcohol based active films for food packaging. *Heliyon* **2021**, *7*, e06550. [CrossRef]
8. DeButts, B.L.; Spivey, C.R.; Barone, J.R. Wheat gluten aggregates as a reinforcement for poly(Vinyl alcohol) films. *ACS Sustain. Chem. Eng.* **2018**, *6*, 2422–2430. [CrossRef]
9. García-Hernández, A.; Morales-Sánchez, E.; Berdeja-Martínez, B.; Escamilla-García, M.; Salgado-Cruz Ma Rentería-Ortega, M.; Farrera-Rebollo, R.; Vega-Cuellar, M.; Calderón-Domínguez, G. Pva-based electrospun biomembranes with hydrolyzed collagen and ethanolic extract of hypericum perforatum for potential use as wound dressing: Fabrication and characterization. *Polymers* **2022**, *14*, 1981. [CrossRef]
10. Damayanti, R.; Tamrin; Alfian, Z.; Eddyanto, E. Preparation film gelatin PVA/gelatin and characterization mechanical properties. *AIP Conf. Proc.* **2021**, *2342*, 060004. [CrossRef]
11. Abedi-Firoozjah, R.; Chabook, N.; Rostami, O.; Heydari, M.; Kolahdouz-Nasiri, A.; Javanmardi, F.; Abdolmaleki, K.; Mousavi Khaneghah, A. PVA/starch films: An updated review of their preparation, characterization, and diverse applications in the food industry. *Polym. Test.* **2023**, *118*, 107903. [CrossRef]
12. Vázquez-Luna, A.; Santiago, M.; Rivadeneyra-Domínguez, E.; Díaz-Sobac, R. Películas comestibles a base de almidón nanoestructurado como material de barrera a la humedad. *CienciaUAT* **2019**, *13*, 152. [CrossRef]
13. Dey, D.; Dharini, V.; Periyar Selvam, S.; Rotimi Sadiku, E.; Mahesh Kumar, M.; Jayaramudu, J.; Nath Gupta, U. Physical, antifungal, and biodegradable properties of cellulose nanocrystals and chitosan nanoparticles for food packaging application. *Mater. Today Proc.* **2021**, *38*, 860–869. [CrossRef]
14. Souza VG, L.; Alves, M.M.; Santos, C.F.; Ribeiro IA, C.; Rodrigues, C.; Coelhoso, I.; Fernando, A.L. Biodegradable chitosan films with zno nanoparticles synthesized using food industry by-products—Production and characterization. *Coatings* **2021**, *11*, 646. [CrossRef]
15. Sayadi, M.; Mojaddar Langroodi, A.; Amiri, S.; Radi, M. Effect of nanocomposite alginate-based film incorporated with cumin essential oil and TiO 2 nanoparticles on chemical, microbial, and sensory properties of fresh meat/beef. *Food Sci. Nutr.* **2022**, *10*, 1401–1413. [CrossRef] [PubMed]
16. Couto, C.; Almeida, A. Metallic nanoparticles in the food sector: A mini-review. *Foods* **2022**, *11*, 402. [CrossRef]
17. Gómez León, M.M.; Román Mendoza, L.E.; Castro Basurto, F.V.; Maúrtua Torres, D.J.; Condori, C.; Vivas, D.; Bianchi, A.E.; Paraguay Delgado, F.; Solís Veliz, J.L. Nanopartículas de CuO y su propiedad antimicrobiana en cepas intrahospitalarias. *Rev. Colomb. De Química* **2017**, *46*, 28–36. [CrossRef]
18. Fernández-León, K.J.; Rodríguez-Díaz, J.A.; Reyes-Espinosa, L.; Duquesne-Alderete, A.; Solenzal-Valdivia, Y.O.; Rives-Quintero, A.; Hernández-García, J.E. Comparison of in vitro anti- Staphylococcus aureus activity of eight antibiotics and four dilutions of propolis. *J. Selva Andin. Res. Soc.* **2022**, *13*, 35–48. [CrossRef]
19. Abud Blanco, K.; Bustos Blanco, L.; Covo Morales, E.; Fang Mercado, L.C. Actividad antimicrobiana in vitro de compuestos fenólicos sulfonados en cavidad oral. *Cienc. Y Salud Virtual* **2015**, *7*, 53. [CrossRef]
20. Heredia-Castro, P.Y.; García-Baldenegro, C.V.; Santos-Espinosa, A.; de Jesús Tolano-Villaverde, J.; Manzanarez-Quin, C.G.; Valdez-Domínguez, R.D.; Ibarra-Zazueta, C.; Osuna-Chávez, R.F.; Rueda-Puente, E.O.; Hernández-Moreno, C.G.; et al. Perfil fitoquímico, actividad antimicrobiana y antioxidante de extractos de Gnaphalium oxyphyllum y Euphorbia maculata nativas de Sonora, México. *Rev. Mex. Cienc. Pecu.* **2022**, *13*, 928–942. [CrossRef]
21. Correa, M.L.; Alfaro, M.E.; Carballo, S.M.; Ureña, Y.C.; Vega-Baudrit, J. Estudio preliminar en la obtención de compuestos híbridos de quitosano y polifenoles derivados de lignina a partir de subproductos agropecuarios y pesquería de camarón. *Rev. Científica* **2017**, *27*, 33–43. [CrossRef]
22. Kong, M.; Chen, X.G.; Xing, K.; Park, H.J. Antimicrobial properties of chitosan and mode of action: A state of the art review. *Int. J. Food Microbiol.* **2010**, *144*, 51–63. [CrossRef] [PubMed]
23. Massana Roquero, D.; Bollella, P.; Katz, E.; Melman, A. Controlling porosity of calcium alginate hydrogels by interpenetrating polyvinyl alcohol–diboronate polymer network. *ACS Appl. Polym. Mater.* **2021**, *3*, 1499–1507. [CrossRef]
24. Luque, G.C.; Picchio, M.L.; Martins, A.P.S.; Dominguez-Alfaro, A.; Tomé, L.C.; Mecerreyes, D.; Minari, R.J. Elastic and thermoreversible iongels by supramolecular pva/phenol interactions. *Macromol. Biosci.* **2020**, *20*, 2000119. [CrossRef] [PubMed]
25. Patil, S.; Bharimalla, A.K.; Mahapatra, A.; Dhakane-Lad, J.; Arputharaj, A.; Kumar, M.; Raja AS, M.; Kambli, N. Effect of polymer blending on mechanical and barrier properties of starch-polyvinyl alcohol based biodegradable composite films. *Food Biosci.* **2021**, *44*, 101352. [CrossRef]
26. Olewnik-Kruszkowska, E.; Gierszewska, M.; Jakubowska, E.; Tarach, I.; Sedlarik, V.; Pummerova, M. Antibacterial films based on pva and pva–chitosan modified with poly(Hexamethylene guanidine). *Polymers* **2019**, *11*, 2093. [CrossRef]
27. Abral, H.; Ikhsan, M.; Rahmadiawan, D.; Handayani, D.; Sandrawati, N.; Sugiarti, E.; Muslimin, A.N. Anti-UV, antibacterial, strong, and high thermal resistant polyvinyl alcohol/Uncaria gambir extract biocomposite film. *J. Mater. Res. Technol.* **2022**, *17*, 2193–2202. [CrossRef]
28. Aslam, M.; Raza, Z.A.; Siddique, A. Fabrication and chemo-physical characterization of CuO/chitosan nanocomposite-mediated tricomponent PVA films. *Polym. Bull.* **2021**, *78*, 1955–1965. [CrossRef]
29. Mansur, H.S.; Oréfice, R.L.; Mansur, AA.P. Characterization of poly(Vinyl alcohol)/poly(Ethylene glycol) hydrogels and PVA-derived hybrids by small-angle X-ray scattering and FTIR spectroscopy. *Polymer* **2004**, *45*, 7193–7202. [CrossRef]

30. Okahisa, Y.; Matsuoka, K.; Yamada, K.; Wataoka, I. Comparison of polyvinyl alcohol films reinforced with cellulose nanofibers derived from oil palm by impregnating and casting methods. *Carbohydr. Polym.* **2020**, *250*, 116907. [CrossRef]
31. Yasmeen, S.; Kabiraz, M.; Saha, B.; Qadir, M.; Gafur, M.; Masum, S. Chromium (Vi) ions removal from tannery effluent using chitosan-microcrystalline cellulose composite as adsorbent. *Int. Res. J. Pure Appl. Chem.* **2016**, *10*, 1–14. [CrossRef]
32. Drabczyk, A.; Kudłacik-Kramarczyk, S.; Głąb, M.; Kędzierska, M.; Jaromin, A.; Mierzwiński, D.; Tyliszczak, B. Physicochemical investigations of chitosan-based hydrogels containing aloe vera designed for biomedical use. *Materials* **2020**, *13*, 3073. [CrossRef]
33. Abu Elella, M.H.; Shalan, A.E.; Sabaa, M.W.; Mohamed, R.R. One-pot green synthesis of antimicrobial chitosan derivative nanocomposites to control foodborne pathogens. *RSC Adv.* **2022**, *12*, 1095–1104. [CrossRef] [PubMed]
34. Tanwar, R.; Gupta, V.; Kumar, P.; Kumar, A.; Singh, S.; Gaikwad, K.K. Development and characterization of PVA-starch incorporated with coconut shell extract and sepiolite clay as an antioxidant film for active food packaging applications. *Int. J. Biol. Macromol.* **2021**, *185*, 451–461. [CrossRef] [PubMed]
35. Gasti, T.; Hiremani, V.D.; Kesti, S.S.; Vanjeri, V.N.; Goudar, N.; Masti, S.P.; Thimmappa, S.C.; Chougale, R.B. Physicochemical and antibacterial evaluation of poly (Vinyl alcohol)/guar gum/silver nanocomposite films for food packaging applications. *J. Polym. Environ.* **2021**, *29*, 3347–3363. [CrossRef]
36. Yang, W.; Ding, H.; Qi, G.; Li, C.; Xu, P.; Zheng, T.; Zhu, X.; Kenny, J.M.; Puglia, D.; Ma, P. Highly transparent PVA/nanolignin composite films with excellent UV shielding, antibacterial and antioxidant performance. *React. Funct. Polym.* **2021**, *162*, 104873. [CrossRef]
37. Remiš, T.; Bělský, P.; Kovářík, T.; Kadlec, J.; Ghafouri Azar, M.; Medlín, R.; Vavruňková, V.; Deshmukh, K.; Sadasivuni, K.K. Study on structure, thermal behavior, and viscoelastic properties of nanodiamond-reinforced poly (Vinyl alcohol) nanocomposites. *Polymers* **2021**, *13*, 1426. [CrossRef] [PubMed]
38. Sánchez-Silva, L.; Víctor-Román, S.; Romero, A.; Gracia, I.; Valverde, J.L. Tailor-made aerogels based on carbon nanofibers by freeze-drying. *Sci. Adv. Mater.* **2014**, *6*, 665–673. [CrossRef]
39. Bueno, J.N.N.; Corradini, E.; de Souza, P.R.; Marques, V.d.S.; Radovanovic, E.; Muniz, E.C. Films based on mixtures of zein, chitosan, and PVA: Development with perspectives for food packaging application. *Polym. Test.* **2021**, *101*, 107279. [CrossRef]
40. Tripathi, R.M.; Pudake, R.N.; Shrivastav, B.R.; Shrivastav, A. Antibacterial activity of poly (Vinyl alcohol)—Biogenic silver nanocomposite film for food packaging material. *Adv. Nat. Sci. Nanosci. Nanotechnol.* **2018**, *9*, 025020. [CrossRef]
41. Ali, H.; Tiama, T.M.; Ismail, A.M. New and efficient NiO/chitosan/polyvinyl alcohol nanocomposites as antibacterial and dye adsorptive films. *Int. J. Biol. Macromol.* **2021**, *186*, 278–288. [CrossRef] [PubMed]
42. Martínez Ferrer, M. Uso de Nuevos Materiales Antimicrobianos Sostenibles Para Envasado Alimentario [Proyecto/Trabajo fin de Carrera/Grado, Universitat Politècnica de València]. 2022. Available online: https://riunet.upv.es/handle/10251/185512 (accessed on 1 July 2022).
43. Montes Hernández, A.I.; Oropeza González, R.A.; Padrón Pereira, C.A.; Araya Quesada, Y.M.; Wexler Goering, L.M.; Cubero Castillo, E.M. Películas biodegradables con propiedades bioactivas. *Rev. Venez. Cienc. Y Tecnol. Aliment.* **2017**, *8*, 057–089. Available online: https://kerwa.ucr.ac.cr/handle/10669/79335 (accessed on 3 July 2020).
44. Ucak, I.; Khalily, R.; Carrillo, C.; Tomasevic, I.; Barba, F.J. Potential of propolis extract as a natural antioxidant and antimicrobial in gelatin films applied to rainbow trout (*Oncorhynchus mykiss*) fillets. *Foods* **2020**, *9*, 1584. [CrossRef] [PubMed]
45. Qian, W.; Liu, M.; Fu, Y.; Zhang, J.; Liu, W.; Li, J.; Li, X.; Li, Y.; Wang, T. Antimicrobial mechanism of luteolin against Staphylococcus aureus and Listeria monocytogenes and its antibiofilm properties. *Microb. Pathog.* **2020**, *142*, 104056. [CrossRef] [PubMed]

Disclaimer/Publisher's Note: The statements, opinions and data contained in all publications are solely those of the individual author(s) and contributor(s) and not of MDPI and/or the editor(s). MDPI and/or the editor(s) disclaim responsibility for any injury to people or property resulting from any ideas, methods, instructions or products referred to in the content.

Proceeding Paper

Food Packaging Film Preparation: From Conventional to Biodegradable and Green Fabrication [†]

Omayra B. Ferreiro [1,2,*] and Magna Monteiro [2]

1. Facultad de Ciencias Químicas, Universidad Nacional de Asunción, San Lorenzo 1055, Paraguay
2. Polytechnic School, National University of Asuncion, Mcal. Estigarribia km 11, San Lorenzo 2111, Paraguay; mmonteiro@pol.una.py
* Correspondence: oferreiro@qui.una.py
† Presented at the 1st International Conference of the Red CYTED ENVABIO100 "Obtaining 100% Natural Biodegradable Films for the Food Industry", San Lorenzo, Paraguay, 14–16 November 2022.

Abstract: It is undeniable that suitable packaging will extend the shelf life of the food. The packaging industry has had to renew and innovate in a world where consumers are increasingly environmentally conscious in order to deal with the impact of the production of petroleum-derived plastics and the management of the waste generated by them. In this way, the use of biopolymers has been proposed, mainly those produced from renewable sources and with biodegradability and/or compostability properties. However, these types of materials are more expensive and do not have the same performance as petroleum-derived materials. Besides, the technologies for film preparation are not adapted for these materials. Therefore, new technologies must be studied and implemented to make the packaging industry a sustainable industry. Recently, non-solvent phase inversion (NIPS) and electrospinning techniques, which are widely used for membrane fabrication, have been proposed for the fabrication of films for food packaging applications from biopolymers and green solvents.

Keywords: biopolymers; films; green chemistry; bio-based polymers

Citation: Ferreiro, O.B.; Monteiro, M. Food Packaging Film Preparation: From Conventional to Biodegradable and Green Fabrication. *Biol. Life Sci. Forum* **2023**, *28*, 11. https://doi.org/10.3390/blsf2023028011

Academic Editor: Mary Lopretty

Published: 14 November 2023

Copyright: © 2023 by the authors. Licensee MDPI, Basel, Switzerland. This article is an open access article distributed under the terms and conditions of the Creative Commons Attribution (CC BY) license (https://creativecommons.org/licenses/by/4.0/).

1. Introduction

Packaging was identified as an important stage of the food supply chain since it is responsible for food safety by extending the life of the food and maintaining its organoleptic characteristics until the final consumer. The goal of the food packaging industry is to provide a quality product as well as safe transportation and distribution to the last stage of the supply chain (the consumption stage) at the lowest possible packaging cost [1]. Consumers have been one of the growth drivers of the food packaging market due to their interest in less processed foods as well as their awareness and expectations regarding the environmental sustainability aspects of food packaging [2–4].

The packaging material depends on many aspects, such as the type of food, industrial processing, shelf-life, the means of transport as well as the distribution and storage requirements [5]. These materials will be in contact with food, so they must be safe and special care must be taken to avoid the eventual migration of undesired constituents into the food. The development of new materials for food packaging is a challenge that concerns researchers together with the food industry to meet consumer expectations as well as avoid socioeconomic problems. Among these socioeconomic problems, it stands out that suitable food packaging could contribute to reducing food loss and waste [6]. It is estimated that by stopping food waste it is possible to save enough food to feed 2 billion hungry people [7]. On the other hand, there is a growing interest in reducing the generation of waste derived from the use of packaging, mainly those made of synthetic plastic.

Conventional plastics are typically made from petrochemical processes and are considered synthetic plastics. These plastics are usually non-biodegradable, which greatly contributes to pollution. For this reason, biodegradable materials have been used as a substitute, mainly for single-use plastics (SUP) [5,8–10]. SUP are widely used in food industry packaging, resulting in a great volume of waste generated [8,11]. In this way, the food packaging industry, which is responsible for over 40% of plastic waste, has promoted the search for new strategies [4]. Furthermore, environmental and regulatory aspects have also contributed to the increase of new eco-friendly materials researched for food packaging towards a green economy [12]. By 2030, the European Commission expects all plastic packaging to be reusable or recyclable [13]. This will contribute to a sustainable packaging industry.

Bio-based plastics are one of the most explored strategies aimed at developing sustainable food packaging materials [3,4,9,14]. However, these materials are not necessarily biodegradable or compostable and need to be recycled. For many years, recycling has been a strategy to reduce the volume of plastic waste and has an environmental benefit since it allows to reduce the use of virgin materials in plastic production [15]. However, it is often impracticable for food packaging due to the possibility of contamination by residual products and particles [5]. Furthermore, it has been reported that recycled food packaging could not be safe due to the increase in potentially hazardous chemical levels in the packaging that could migrate into the food [16]. Biodegradable and compostable packaging are of interest when it is not possible to reduce, reuse or recycle; however, more research and development are needed to satisfy the bio-packaging market [6,17]. It is also worth mentioning that most of the costs involved in manufacturing food packaging are attributable to raw materials, mainly when bio-based plastics and biodegradable materials are considered [5,12,18,19]. This way, it has encouraged the development of technology for obtaining eco-friendly, cheaper and degradable materials from organic waste streams (such as underutilized byproducts from food processing industries and agro-industrial biomass, among others) toward a circular economy not competing with food usage of agricultural resources [6,14,20].

The food packaging industry could contribute to fulfilling some of the United Nations' Sustainable Development Goals by using renewable resources or recycled materials and implementing innovations in the design and production of biodegradable, compostable and recyclable packaging materials [21,22]. However, even meeting these requirements and necessities for being used in the food industry, the materials must be economically competitive with the conventional plastics to be viable to survive in the market and to stimulate the development of this kind of material [4,5].

2. Films for Food Packaging

Polymeric films act as barriers by controlling water vapor and atmospheric gases permeation through them and are, therefore, a popular choice for food packaging purposes [23]. The food packaging films market was worth USD 49.8 billion in 2021 and is projected to reach USD 72.3 billion by 2027 [12]. Multilayer films combine several types of plastic consisting of 3–9 layers and have become important in the food packaging for developing high-performance food packaging in a cost-effective manner [5,24,25].

The most suitable material for film preparation will depend on the product characteristics, but in any case, any undesirable migration from the film to the food must be avoided, which means that the material must be safe [26]. Generally, it is important to know the characteristics of gas permeation, such as oxygen and carbon dioxide, or the exchange rate of these gases through the film. Furthermore, it is important to also consider the permeation of moisture, flavor, and other volatile compounds as well as the selectivity of these gases or vapors. Films with a low moisture permeability are often desired [3,5].

The gas and vapor permeation will depend on the environmental conditions in which the product will be kept, such as humidity, temperature, oxygen, light (which can induce degradation reactions during storage), among others. The mechanical resistance of the film will allow safe handling and transportation of the product through the supply chain. The material must also be easy to process and have migration limits according to the regulations [5,27].

The modification in the gas atmosphere near food products during storage has a great influence on their shelf life [28]. In this way a technology known as modified atmosphere packaging (MAP) has been proposed, in which the atmosphere inside the food package is modified by altering the gas composition or by removing it (vacuum packaging). For fresh products, the goal is to create a balance within the package where the respiratory activity of a product is as low as possible by maintaining a low concentration of oxygen and/or a high concentration of carbon dioxide [29]. However, there are very few materials that are sufficiently permeable to match the respiration rate of fruits and vegetables and each material has to be optimized for specific demands [26,30]. Modifying the atmosphere of packaged foods can improve their visual appearance, texture, and nutritional appeal, and it is considered minimal processing compared to applying food chemicals, preservatives, or stabilizers since it inhibits the microbial growth [28,29].

When antioxidants, nutraceutical or antimicrobial agents, enzymes, oxygen/ethylene carriers, flavor delivery or absorption systems are incorporated into the film matrix or are applied as a coating, the system is considered as active or bioactive food packaging in which the product, package and package environment interact to provide a positive effect on the food [17,31–34]. All active packaging technologies include some physical, chemical, or biological action to generate interactions between the packaging, the product and the space left between the product and the packaging, aiming to increase the shelf life of the food by a controlled absorbing or releasing of active ingredients or by scavenging undesirable substances [9]. The active compounds can be added to or be naturally present in the raw materials used for the film preparation, and it is important to know the toxicity of these compounds, their action mechanisms and their stability [32].

3. Eco-Friendly Materials for Film Preparation

In 2021, 390.7 million tons of plastics were produced worldwide, and 90.2% of these came from fossil sources, while post-consumer recycled plastics and bio-based/bio-attributed plastics accounted for 8.3% and 1.5% of world production, respectively. This year, packaging, building and construction applications were the two largest world plastics markets. Particularly in Europe, more than 50 million tons were produced, and packaging represented 39% [35]. For packaging applications, the most commonly used polymers are low-density polyethylene (LDPE), high-density polyethylene (HDPE), polypropylene (PP), polytetrafluoroethylene (PTFE), nylon (polyamide), polyethylene terephthalate (PET), polyesters, polyvinyl chloride (PVC), polystyrene (PS) and ethylene vinyl acetate (EVA) [17,29,35].

Most of these polymers have a low cost and appropriate physical, mechanical and transport properties to be used in food packaging. However, they are petroleum-derived polymers and have low or no biodegradability [29]. Degradability is the material ability to break down into carbon dioxide, methane, water, inorganic components, and biomass, and it could be done chemically or biologically. Biodegradability is performed using microorganisms, such as fungi and bacteria, to achieve the breakdown of matter into lower molecular-weight products that can then be used by other organisms until the complete decomposition of matter. It is worth mentioning that the rate of degradation is highly dependent on the chemical structure [36].

There has been a growing interest in reducing the environmental impact of plastics; therefore, new materials have been studied for food packaging, such as bio-based polymers and biodegradable polymers. Biopolymers include these two types of polymers and are polymers that can be extracted from biomass (such as cellulose or starch), synthesized from bio-derived monomers (such as Bio-PP or polylactic acid), and produced from microorganisms (such as polyhydroxy-alkanoates) [21,29,36,37].

These materials are being designed to replace synthetic plastic materials with the goal of achieving a minimal carbon footprint, high recycling value, or complete biodegradability or compostability. Although not all bio-based polymers are inherently biodegradable, some of them could exhibit antioxidant and antimicrobial activity and biocompatibility, among other positive effects [21,36,38]. Moreover, if these materials come from renewable waste streams or biomass sources that are not competent with food and agricultural resources, sustainable development is achieved by promoting a circular bioeconomy [14,21,36].

As previously mentioned, polymers for food packaging must be thermally stable, flexible and have a good barrier to gases and chemicals, which will depend mainly on the packaging matrix. However, most of the biopolymers used for food packaging have been reported to have poor mechanical or barrier properties toward moisture and water vapor compared to synthetic polymers [37,39]. Furthermore, while the technologies to produce synthetic plastics are widely established, the technologies to produce bioplastics lack comparable scalability and productivity. These factors have delayed the widening applicability of biopolymers in food packaging [21,37].

The food packaging industry has been working on innovative solutions for improving the barrier performance, mechanical strength, and thermal stability of the packaging and, consequently, extending the food shelf life [31,40–42]. Among the strategies to improve the characteristics of biodegradable polymers have been suggested: the use of nanotechnology by the incorporation of nanofillers such as nanoparticles to modify or control the permeability or the release of active ingredients or to provide antioxidant, antibacterial, antifungal or antimicrobial properties [9,31,33,37,42–44]; the application of a surface treatment such as coating using a good film-forming [41,45] or inducing a crosslinking [46]; and the blending of biopolymers, which should be compatible [5,47].

Among the biopolymers most reported in the literature with a great potential to substitute synthetic polymers in the food packaging industry can be cited: starch, chitosan, polylactic acid (PLA), and polyhydroxyalkanoates (PHA) [5,48,49]. PLA is the biopolymer with the greatest potential to replace petroleum-based polymers (such as polystyrene (PS) and polypropylene (PP)) in packaging applications due to its excellent barrier properties [38,39]. Oriented PLA (OPLA) showed to be a good film for tomatoes and other breathable products because of the matching of the oxygen and carbon dioxide exchange with the respiration rate of these products [27]. Table 1 summarizes some of the advantages and limitations for the biopolymers more commonly used and reported in the literature.

Table 1. Advantages and limitations of biopolymers for food packaging applications.

Biopolymer	Advantages	Limitations	Ref
Chitosan	<ul><li>Availability</li><li>Nontoxic</li><li>Easy to form films</li><li>Selectivity to carbon dioxide</li><li>Oxygen permeability</li><li>Antibacterial, antifungal, and mechanical properties</li></ul>	<ul><li>High sensitivity to water afecting its mechanical stability</li></ul>	[5,50]
Starch	<ul><li>Low Price</li><li>Zero toxicity</li><li>High degradability</li><li>Easy availability</li><li>Already being commercialized</li></ul>	<ul><li>Poor resistance to humidity</li><li>Poor thermal processability</li><li>Poor mechanical resistance</li><li>It is not stable to heat</li></ul>	[51]
PLA	<ul><li>The existing technology for the manufacture of films can be used</li><li>It has properties similar to polymers of fossil origin (oxygen transfer; strength and stability)</li></ul>	<ul><li>Fragile</li></ul>	[9,38]

Table 1. *Cont.*

Biopolymer	Advantages	Limitations	Ref
PHA	• Not toxic • Insoluble in water • Biodegradable/Biocompatible/Renewable • Good UV resistance • High tensile strength	• Cost of raw material when pure glucose is used • Low barrier properties	[9,44,52]

4. Films Preparation Techniques

Although biopolymers do not present the same performance for food packaging applications, another factor that has limited their widening application is that the current film preparation techniques are not always suitable for these materials. Among the techniques for synthetic polymers that have been tested for biopolymers is blown film extrusion. This technique has undergone extensive industrial development for synthetic polymers. However, it presents some limitations for biopolymers, as will be discussed below. Solution casting is the most reported technique for biopolymer film preparation, although this technique is difficult to implement on an industrial scale.

On the other hand, membrane preparation technology for separation processes is quite developed. In this way, it is possible to take advantage of this area to design biopolymer-based food packaging films. Non-solvent-induced phase inversion (NIPS) is a conventional method for fabricating polymeric membranes, while electrospinning is an emergent technique. Both techniques could be used to prepare films for food packaging.

The blown-extrusion technique allows for the production of polymeric films of a variety of thicknesses on a large scale due to its low cost, continuity and simplicity of operation. Multilayer films can also be produced by this technique. PLA films were successfully produced by this technique by adding nanoparticles of MgO, which enhanced the film plasticity [43]. Karkhanis et al. [53] prepared transparent PLA films containing cellulose nanocrystals with high potential to be used for food packaging due to the enhanced barrier performance obtained. Bilayer biodegradable films were also prepared by the co-extrusion of PLA and Bio-flex® in blown-extruder equipment, showing that it is possible to obtain multilayer films in a single step [54].

However, blown-extrusion processing requires a high melt viscosity resin, which limits the application of other biopolymers different from PLA, and it is necessary to blend the biopolymers with other polymers to improve their processability by modifying the rheological properties of the blend [5,29,51].

Solution casting is the most popular technique to prepare biopolymeric films on a laboratory scale because of its simplicity [5,51]. Besides, this technique allows for a better crosslinking between two or more blended polymers [45]. However, this technique has a high energy consumption for solvent evaporation (commonly, water), which has hindered its expansion to an industrial scale [51]. Furthermore, the incorporation of nanoparticles into the film is also limited due to the difficulties related to their uniform dispersion in the film [5]. This could be overcome using tip sonication; this way, Manikandan, Pakshirajan and Pugazhenthi [44] prepared PHA films with graphene nanoplatelets, which were uniformly distributed in the PHA matrix. On the other hand, Ochoa-Yepes et al. [55] reported better mechanical and lower moisture content and water vapor permeability for the starch films prepared by the extrusion/thermocompression process than the ones prepared by solution casting.

The NIPS technique for the preparation of biodegradable films has recently proposed by Liu et al. [56]. The authors prepared PLA films by NIPS for pork meat packaging to extend the shelf life of the food. Although the authors used N-methylpyrrolidone as a solvent, green solvents could be explored as an alternative to a sustainable film preparation. NIPS is a technique widely used for polymeric membrane preparation. The membrane industry is also looking for eco-friendly alternatives toward a sustainable membrane fabrication process. In this way, biopolymers and green solvents have been proposed [57,58]. On the other hand, electrospinning technology has been widely investigated for the fabrication of nanofibrous membranes for water treatment [59] and more recently for the fabrication of food packaging materials [60]. PLA-based nanomaterials as well as active and intelligent packaging materials have been fabricated for food packaging applications by electrospinning with expected structures and enhanced barrier, mechanical, and thermal properties [39].

5. Final Considerations

The food packaging film industry is directed towards sustainable development; for this reason, it requires raw materials obtained from renewable or recycled sources, more efficient and green production, and proper waste management.

Author Contributions: All authors contributed equally. All authors have read and agreed to the published version of the manuscript.

Funding: This research was funded by Red Cyted ENVABIO100 121RT0108.

Institutional Review Board Statement: Not applicable.

Informed Consent Statement: Not applicable.

Data Availability Statement: Data sharing is not applicable to this article.

Conflicts of Interest: The authors declare no conflict of interest.

References

1. González, A.; Contreras, C.B.; Alvarez Igarzabal, C.I.; Strumia, M.C. Study of the structure/property relationship of nanomaterials for development of novel food packaging. In *Food Packaging*; Academic Press: Cambridge, MA, USA, 2017; pp. 265–294. [CrossRef]
2. Chirilli, C.; Molino, M.; Torri, L. Consumers' Awareness, Behavior and Expectations for Food Packaging Environmental Sustainability: Influence of Socio-Demographic Characteristics. *Foods* **2022**, *11*, 2388. [CrossRef] [PubMed]
3. Mendes, A.C.; Pedersen, G.A. Perspectives on sustainable food packaging: Is bio-based plastics a solution? *Trends Food Sci. Technol.* **2021**, *112*, 839–846. [CrossRef]
4. Tyagi, P.; Salem, K.S.; Hubbe, M.A.; Pal, L. Advances in barrier coatings and film technologies for achieving sustainable packaging of food products—A review. *Trends Food Sci. Technol.* **2021**, *115*, 461–485. [CrossRef]
5. Kumari, S.V.G.; Pakshirajan, K.; Pugazhenthi, G. Recent advances and future prospects of cellulose, starch, chitosan, polylactic acid and polyhydroxyalkanoates for sustainable food packaging applications. *Int. J. Biol. Macromol.* **2022**, *221*, 163–182. [CrossRef] [PubMed]
6. Guillard, V.; Gaucel, S.; Fornaciari, C.; Angellier-Coussy, H.; Buche, P.; Gontard, N. The Next Generation of Sustainable Food Packaging to Preserve Our Environment in a Circular Economy Context. *Front. Nutr.* **2018**, *5*, 121. [CrossRef]
7. World Food Program USA. Available online: https://www.wfpusa.org/articles/how-food-waste-affects-world-hunger/ (accessed on 31 January 2023).
8. Dey, A.; Dhumal, C.V.; Sengupta, P.; Kumar, A.; Pramanik, N.K.; Alam, T. Challenges and possible solutions to mitigate the problems of single-use plastics used for packaging food items: A review. *J. Food Sci. Technol.* **2021**, *58*, 3251–3269. [CrossRef]
9. Jariyasakoolroj, P.; Leelaphiwat, P.; Harnkarnsujarit, N. Advances in research and development of bioplastic for food packaging. *J. Sci. Food Agric.* **2020**, *100*, 5032–5045. [CrossRef]
10. Shahabi-Ghahfarrokhi, I.; Almasi, H.; Babaei-Ghazvini, A. Characteristics of biopolymers from natural resources. In *Processing and Development of Polysaccharide-Based Biopolymers for Packaging Applications*; Elsevier: Amsterdam, The Netherlands, 2020; pp. 49–95. [CrossRef]
11. Gallego-Schmid, A.; Mendoza, J.M.F.; Azapagic, A. Environmental impacts of takeaway food containers. *J. Clean. Prod.* **2019**, *211*, 417–427. [CrossRef]
12. Markets and Markets. Food Packaging Films Market. Available online: https://www.marketsandmarkets.com/Market-Reports/food-packaging-films-market-155846613.html (accessed on 31 January 2023).
13. European Commission. Questions & Answers: A European Strategy for Plastics. 2018. Available online: https://ec.europa.eu/commission/presscorner/detail/en/MEMO_18_6 (accessed on 31 January 2023).
14. Suresh, S.; Pushparaj, C.; Subramani, R. Recent development in preparation of food packaging films using biopolymers. *Food Res.* **2021**, *5*, 12–22. [CrossRef]
15. Hou, P.; Xu, Y.; Taiebat, M.; Lastoskie, C.; Miller, S.A.; Xu, M. Life cycle assessment of end-of-life treatments for plastic film waste. *J. Clean. Prod.* **2018**, *201*, 1052–1060. [CrossRef]

16. Geueke, B.; Groh, K.; Muncke, J. Food packaging in the circular economy: Overview of chemical safety aspects for commonly used materials. *J. Clean. Prod.* **2018**, *193*, 491–505. [CrossRef]
17. McMillin, K.W. Advancements in meat packaging. *Meat. Sci.* **2017**, *132*, 153–162. [CrossRef]
18. Stoica, M.; Marian Antohi, V.; Laura Zlati, M.; Stoica, D. The financial impact of replacing plastic packaging by biodegradable biopolymers—A smart solution for the food industry. *J. Clean. Prod.* **2020**, *277*. [CrossRef]
19. Rahman, M.H.; Bhoi, P.R. An overview of non-biodegradable bioplastics. *J. Clean. Prod.* **2021**, *294*. [CrossRef]
20. Velázquez, M.E.; Ferreiro, O.B.; Menezes, D.B.; Corrales-Ureña, Y.; Vega-Baudrit, J.R.; Rivaldi, J.D. Nanocellulose Extracted from Paraguayan Residual Agro-Industrial Biomass: Extraction Process, Physicochemical and Morphological Characterization. *Sustainability* **2022**, *14*, 1386. [CrossRef]
21. RameshKumar, S.; Shaiju, P.; O'Connor, K.E.; P, R.B. Bio-based and biodegradable polymers—State-of-the-art, challenges and emerging trends. *Curr. Opin. Green Sustain. Chem.* **2020**, *21*, 75–81. [CrossRef]
22. Green Blue. Addressing The Sustainable Development Goals Through Packaging: SDG 12, Sustainable Production & Consumption. Available online: https://greenblue.org/addressing-the-sustainable-development-goals-through-packaging-sdg-12-sustainable-production-consumption/ (accessed on 31 January 2023).
23. McKeen, L.W. Introduction to Use of Plastics in Food Packaging. In *Plastic Films in Food Packaging*; Plastics Design Library, William Andrew Publishing: Norwich, NY, USA, 2013; pp. 1–15. [CrossRef]
24. Mount, E.M. Coextrusion Equipment for Multilayer Flat Films and Sheets. In *Multilayer Flexible Packaging*; Plastics Design Library, William Andrew Publishing: Norwich, NY, USA, 2016; pp. 99–122. [CrossRef]
25. Yun, X.; Dong, T. Fabrication of high-barrier plastics and its application in food packaging. In *Food Packaging*; Academic Press: Cambridge, MA, USA, 2017; pp. 147–184. [CrossRef]
26. Peinemann, K.-V.; Pereira Nunes, S.; Giorno, L. *Membrane Technology*; Volume 3: Membranes for Food Applications; WILEY-VCH Verlag GmbH & Co. KGaA: Weinheim, Germany, 2010.
27. Siracusa, V.; Rocculi, P.; Romani, S.; Rosa, M.D. Biodegradable polymers for food packaging: A review. *Trends Food Sci. Technol.* **2008**, *19*, 634–643. [CrossRef]
28. Priyadarshi, R.; Deeba, F.; Sauraj; Negi, Y.S. Modified atmosphere packaging development. In *Processing and Development of Polysaccharide-Based Biopolymers for Packaging Applications*; Elsevier: Amsterdam, The Netherlands, 2020; pp. 261–280. [CrossRef]
29. Mangaraj, S.; Yadav, A.; Bal, L.M.; Dash, S.K.; Mahanti, N.K. Application of Biodegradable Polymers in Food Packaging Industry: A Comprehensive Review. *J. Packag. Technol. Res.* **2018**, *3*, 77–96. [CrossRef]
30. Goswami, T.K.; Mangaraj, S. Advances in polymeric materials for modified atmosphere packaging (MAP). In *Multifunctional and Nanoreinforced Polymers for Food Packaging*; Woodhead Publishing: Cambridge, UK, 2011; pp. 163–242.
31. Vasile, C. Polymeric Nanocomposites and Nanocoatings for Food Packaging: A Review. *Materials* **2018**, *11*, 1834. [CrossRef]
32. Tapia-Blácido, D.R.; da Silva Ferreira, M.E.; Aguilar, G.J.; Lemos Costa, D.J. Biodegradable packaging antimicrobial activity. In *Processing and Development of Polysaccharide-Based Biopolymers for Packaging Applications*; Elsevier: Amsterdam, The Netherlands, 2020; pp. 207–238. [CrossRef]
33. Jacob, J.; Thomas, S.; Loganathan, S.; Valapa, R.B. Antioxidant incorporated biopolymer composites for active packaging. In *Processing and Development of Polysaccharide-Based Biopolymers for Packaging Applications*; Elsevier: Amsterdam, The Netherlands, 2020; pp. 239–260. [CrossRef]
34. Antosik, A.K.; Kowalska, U.; Stobinska, M.; Dzieciol, P.; Pieczykolan, M.; Kozlowska, K.; Bartkowiak, A. Development and Characterization of Bioactive Polypropylene Films for Food Packaging Applications. *Polymers* **2021**, *13*, 3478. [CrossRef]
35. Plastics-Europe. Plastics, The Facts 2022. Plastics-Europe. 2022. Available online: https://plasticseurope.org/knowledge-hub/plastics-the-facts-2022 (accessed on 31 January 2023).
36. Garrison, T.F.; Murawski, A.; Quirino, R.L. Bio-Based Polymers with Potential for Biodegradability. *Polymers* **2016**, *8*, 262. [CrossRef] [PubMed]
37. Nesic, A.R.; Seslija, S.I. The influence of nanofillers on physical–chemical properties of polysaccharide-based film intended for food packaging. In *Food Packaging*; Academic Press: Cambridge, MA, USA, 2017; pp. 637–697. [CrossRef]
38. Jacob, J.; Lawal, U.; Thomas, S.; Valapa, R.B. Biobased polymer composite from poly(lactic acid): Processing, fabrication, and characterization for food packaging. In *Processing and Development of Polysaccharide-Based Biopolymers for Packaging Applications*; Elsevier: Amsterdam, The Netherlands, 2020; pp. 97–115. [CrossRef]
39. Wu, J.H.; Hu, T.G.; Wang, H.; Zong, M.H.; Wu, H.; Wen, P. Electrospinning of PLA Nanofibers: Recent Advances and Its Potential Application for Food Packaging. *J. Agric. Food Chem.* **2022**, *70*, 8207–8221. [CrossRef] [PubMed]
40. Han, J.W.; Ruiz-Garcia, L.; Qian, J.P.; Yang, X.T. Food Packaging: A Comprehensive Review and Future Trends. *Compr. Rev. Food Sci. Food Saf.* **2018**, *17*, 860–877. [CrossRef] [PubMed]
41. Wang, W.; Qin, C.; Li, W.; Ge, J.; Feng, C. Improving moisture barrier properties of paper sheets by cellulose stearoyl ester-based coatings. *Carbohydr. Polym.* **2020**, *235*, 115924. [CrossRef]
42. Chi, H.; Song, S.; Luo, M.; Zhang, C.; Li, W.; Li, L.; Qin, Y. Effect of PLA nanocomposite films containing bergamot essential oil, TiO_2 nanoparticles, and Ag nanoparticles on shelf life of mangoes. *Sci. Hortic.* **2019**, *249*, 192–198. [CrossRef]
43. Swaroop, C.; Shukla, M. Development of blown polylactic acid-MgO nanocomposite films for food packaging. *Compos. Part A Appl. Sci. Manuf.* **2019**, *124*, 105482. [CrossRef]

44. Manikandan, N.A.; Pakshirajan, K.; Pugazhenthi, G. Preparation and characterization of environmentally safe and highly biodegradable microbial polyhydroxybutyrate (PHB) based graphene nanocomposites for potential food packaging applications. *Int. J. Biol. Macromol.* **2020**, *154*, 866–877. [CrossRef]
45. Liu, Y.; Ahmed, S.; Sameen, D.E.; Wang, Y.; Lu, R.; Dai, J.; Li, S.; Qin, W. A review of cellulose and its derivatives in biopolymer-based for food packaging application. *Trends Food Sci. Technol.* **2021**, *112*, 532–546. [CrossRef]
46. Wu, H.; Lei, Y.; Lu, J.; Zhu, R.; Xiao, D.; Jiao, C.; Xia, R.; Zhang, Z.; Shen, G.; Liu, Y.; et al. Effect of citric acid induced crosslinking on the structure and properties of potato starch/chitosan composite films. *Food Hydrocoll.* **2019**, *97*, 105208. [CrossRef]
47. Claro, P.I.C.; Neto, A.R.S.; Bibbo, A.C.C.; Mattoso, L.H.C.; Bastos, M.S.R.; Marconcini, J.M. Biodegradable Blends with Potential Use in Packaging: A Comparison of PLA/Chitosan and PLA/Cellulose Acetate Films. *J. Polym. Environ.* **2016**, *24*, 363–371. [CrossRef]
48. Flórez, M.; Cazón, P.; Vázquez, M. Active packaging film of chitosan and Santalum album essential oil: Characterization and application as butter sachet to retard lipid oxidation. *Food Packag. Shelf Life* **2022**, *34*, 100938. [CrossRef]
49. Bhowmik, S.; Agyei, D.; Ali, A. Bioactive chitosan and essential oils in sustainable active food packaging: Recent trends, mechanisms, and applications. *Food Packag. Shelf Life* **2022**, *34*, 100962. [CrossRef]
50. Haghighi, H.; Licciardello, F.; Fava, P.; Siesler, H.W.; Pulvirenti, A. Recent advances on chitosan-based films for sustainable food packaging applications. *Food Packag. Shelf Life* **2020**, *26*, 100551. [CrossRef]
51. do Val Siqueira, L.; Arias, C.I.L.F.; Maniglia, B.C.; Tadini, C.C. Starch-based biodegradable plastics: Methods of production, challenges and future perspectives. *Curr. Opin. Food Sci.* **2021**, *38*, 122–130. [CrossRef]
52. Singh, A.K.; Singh, S.P.; Porwal, P.; Pandey, B.; Srivastava, J.K.; Ansari, M.I.; Chandel, A.K.; Rathore, S.S.; Mala, J. Processes and characterization for biobased polymers from polyhydroxyalkanoates. In *Processing and Development of Polysaccharide-Based Biopolymers for Packaging Applications*; Elsevier: Amsterdam, The Netherlands, 2020; pp. 117–149. [CrossRef]
53. Karkhanis, S.S.; Stark, N.M.; Sabo, R.C.; Matuana, L.M. Water vapor and oxygen barrier properties of extrusion-blown poly(lactic acid)/cellulose nanocrystals nanocomposite films. *Compos. Part A Appl. Sci. Manuf.* **2018**, *114*, 204–211. [CrossRef]
54. Scaffaro, R.; Maio, A.; Gulino, F.E.; Di Salvo, C.; Arcarisi, A. Bilayer biodegradable films prepared by co-extrusion film blowing: Mechanical performance, release kinetics of an antimicrobial agent and hydrolytic degradation. *Compos. Part A Appl. Sci. Manuf.* **2020**, *132*. [CrossRef]
55. Ochoa-Yepes, O.; Di Giogio, L.; Goyanes, S.; Mauri, A.; Fama, L. Influence of process (extrusion/thermo-compression, casting) and lentil protein content on physicochemical properties of starch films. *Carbohydr. Polym.* **2019**, *208*, 221–231. [CrossRef]
56. Liu, W.; Huang, N.; Yang, J.; Peng, L.; Li, J.; Chen, W. Characterization and application of porous polylactic acid films prepared by nonsolvent-induced phase separation method. *Food Chem.* **2022**, *373*, 131525. [CrossRef]
57. Galiano, F.; Briceño, K.; Marino, T.; Molino, A.; Christensen, K.V.; Figoli, A. Advances in biopolymer-based membrane preparation and applications. *J. Membr. Sci.* **2018**, *564*, 562–586. [CrossRef]
58. Naziri Mehrabani, S.A.; Vatanpour, V.; Koyuncu, I. Green solvents in polymeric membrane fabrication: A review. *Sep. Purif. Technol.* **2022**, *298*, 121691. [CrossRef]
59. Mohammad Mahdi, A.S.; Saeed, B.; Fereshteh, M. Electrospun Nanofibrous Membranes for Water Treatment. In *Advances in Membrane Technologies*; Amira, A., Ed.; IntechOpen: Rijeka, Italy, 2020.
60. Zhao, L.; Duan, G.; Zhang, G.; Yang, H.; He, S.; Jiang, S. Electrospun Functional Materials toward Food Packaging Applications: A Review. *Nanomaterials* **2020**, *10*, 150. [CrossRef] [PubMed]

Disclaimer/Publisher's Note: The statements, opinions and data contained in all publications are solely those of the individual author(s) and contributor(s) and not of MDPI and/or the editor(s). MDPI and/or the editor(s) disclaim responsibility for any injury to people or property resulting from any ideas, methods, instructions or products referred to in the content.

Proceeding Paper

Biotransformation of Rice Husk into Phenolic Extracts by Combined Solid Fermentation and Enzymatic Treatment [†]

Maria Inês Dias [1,2], José Pinela [1,2], Tânia C. S. P. Pires [1,2,3], Filipa Mandim [1,2], Maria-Filomena Barreiro [1,2], Lillian Barros [1,2], José Roberto Vega-Baudrit [4], Isabel C. F. R. Ferreira [1,2] and Mary Lopretti [5,*]

1. Centro de Investigação de Montanha (CIMO), Instituto Politécnico de Bragança, Campus de Santa Apolónia, 5300-253 Bragança, Portugal; maria.ines@ipb.pt (M.I.D.); jpinela@ipb.pt (J.P.); tania.pires@ipb.pt (T.C.S.P.P.); filipamandim@ipb.pt (F.M.); barreiro@ipb.pt (M.-F.B.); lillian@ipb.pt (L.B.); iferreira@ipb.pt (I.C.F.R.F.)
2. Laboratório Associado para a Sustentabilidade e Tecnologia em Regiões de Montanha (SusTEC), Instituto Politécnico de Bragança, Campus de Santa Apolónia, 5300-253 Bragança, Portugal
3. Nutrition and Bromatology Group, Department of Analytical Chemistry and Food Science, Faculty of Science, Universidad de Vigo, E-32004 Ourense, Spain
4. Laboratorio Nacional de Nanotecnología-CeNAT-CONARE, San José 1174-1200, Costa Rica; jvegab@gmail.com
5. Centro de Investigaciones Nucleares, Facultad de Ciencias, Universidad de la República, Aigua, Montevideo 4225, Uruguay
* Correspondence: mlopretti@gmail.com
† Presented at the 1st International Conference of the Red CYTED ENVABIO100 "Obtaining 100% Natural Biodegradable Films for the Food Industry", San Lorenzo, Paraguay, 14–16 November 2022.

Abstract: Biotechnology is essential for developing profitable and productive techniques to obtain metabolites. Two technologies can be used: solid or liquid fermentation and enzymatic treatments. In this context, the objective of this work was to evaluate the use of rice husk, a lignocellulosic material, to obtain bioactive compounds by lignin oxidative transformation and demethoxylation, respectively, through enzymatic treatments of *P. chrysosporium* and *G. trabeum*. In the first step, solid fermentation was used to obtain the enzyme Lig. Peroxidase and methoxyl hydrolase were quantified as 80 UE and 50 UE, respectively. This enzyme concentrate was lyophilized and used to prepare an enzymatic consortium (240 UE LigP and 150 UE metH) applied in the second phase of enzymatic treatment. The overall process involved 20 days in the solid fermentation step and 2 h for the enzymatic treatment. The obtained products were characterized by having veratryl alcohol and veratryl aldehyde at contents of 70.4 ± 0.1 and 23.3 ± 0.3 mg/g, respectively. Moreover, the analyzed products did not show cytotoxicity but revealed antioxidant and bacteriostatic activities. No anti-inflammatory activity was detected. In the context of circular economy, the obtained results pointed out the use of combined solid fermentation and enzymatic treatment as a viable strategy to valorize rice husk. The applications of these bioactive compounds presenting bactericidal and bacteriostatic activity and not showing toxicity are very common in medicine, agriculture, and environmental health, among others, and can be incorporated both in free systems and immobilized in spheres, capsules or biopolymer films, which is an important input for obtaining functionalized materials that are in high demand today.

Keywords: solid fermentation; enzymatic treatment; bioactive phenolic compounds

1. Introduction

Fungi represent a good substrate transformation tool, such as rice husk residues, which use cellulose, polyphenols, and hemicelluloses as nutrients for their metabolic growth and development. The substrates used are transformed to obtain compounds with properties due to the biological actions of some of their metabolites. Antioxidant, hypocholesterolemic, hypoglycemic, antibacterial, immunomodulatory, anticancer, and regulatory effects of the cardiovascular and antiviral system have been described [1], the last three being the most relevant. These properties have been attributed to polysaccharides and triterpenoidal compounds, secondary metabolites of some fungi. Many of these compounds are currently being used to produce commercial medicines [2]. However, preliminary research on basidiomycetes, specifically on Shiitake, has shown that these compounds' proportions vary with the fungus's stage and the medium in which it is grown. These skills are presented for fungi such as *Lentinula edodes* (Shitake), *Pleurotus*, *Trametes*, and *Phanaerochaetes*.

It is known that both mycelium and depleted media, as well as different extracts exhibit antibacterial properties, with different effectiveness against Gram-positive compared to Gram-negative bacteria, as is the case for activity against pathogens such as *Bacillus megaterium*, *Streptococcus pyogenes*, and *Staphylococcus aureus*).

Not only intrametabolites present these actions since an exopolymer, a glycoprotein, isolated from Shiitake, when administered in a dose of 200 mg/kg body weight, reduces the plasma level of total cholesterol by 25.1%, while the triglyceride level drops by 44.5%.

Among the compounds that can be obtained are β-glucans, which are non-cellulosic polysaccharides consisting of glucose units linked by glycosidic bonds and with β-1-3 or β-1-6 branches. They are isolated mainly from the fungal cell wall but can also be excreted into the environment. They have immunostimulatory and anticancer activities, and anti-infective, hypocholesterolemic, hypoglycemic, anti-inflammatory, and analgesic properties.

Polyketides are a diverse family of natural products with various pharmacological activities and properties, including antibiotic, antifungal, cytostatic, hypocholesterolemic, antiparasitic, and animal growth and insecticide promotion. Within polyketides statins, polyketides are not aromatic, which was shown to inhibit cholesterol synthesis since they are inhibitors of 3-hydroxy-3-methyl-glutaryl coenzyme A reductase (HMG-CoA), the first enzyme involved in cholesterol biosynthesis.

In addition to the constitutive compounds present in their growth, fungi have the property of transforming the polyphenols present in the substrate where they grow (ligno-cellulosic substrates), obtaining functional phenolic units from that catabolism.

These structures derived from complex polyphenols are a source of few natural phenolic aromatic structures. Polyphenols produced by fungi can be depolymerized and oxidized by acquiring new functional groups, such as aldehydes and acids, with various uses. Today, some fungi have very specific activities, such as hydrolyzing the methoxyl of thephenolic ring in position 3 and position 5, which allows for the ability to obtain a greater quantity of building block-type products to re-engineer products. Within the activities of these phenols, the germicidal activity of some fungi and bacteria can allow for the development of products such as polyurethanes with germicidal activity and phenolic extracts of fungicidal activity.

In the present work, phenolic extracts from rice husk production as a renewable substrate were investigated using a combined system of solid fermentation and enzymatic treatment that allows functional compounds (FC) to be obtained in a standardized way, with the aim of a future scale-up.

2. Materials and Methods

2.1. Rice husk Samples

The rice (*Oryza sativa*, Olimar and Gurí) husk was provided by the Uruguayan rice company COOPAR S.A., Lascano, Rocha Departament 15, 27300.

2.2. Strains Obtainment

Phanerochaete chrysosporium and *Gloeophyllum trabeum* strains were obtained from the TNA Laboratory in Biochemistry and Biotechnology of CIN, Faculty of Sciences of the University of the Republic, Uruguay. They were kept in potato dextrose agar (PDA) medium and incubated at 22 ± 2 °C for 5–8 days. For liquid fermentation, GPEL broth (glucose (20 g/L), extract of yeast (2.5 g/L), peptone (2.5 g/L)) adjusted to a pH of 4.5 with 0.1 N HCL and 0.1 N NaOH was used.

2.3. Strain Obtainment

2.3.1. Inoculation and Fermentation

For the pre-inoculum preparation, 200 mL of GPEL broth was poured into 500 mL flasks and then inoculated with 4 mm paper discs obtained from PDA boxes previously invaded with the prepared mycelium. The incubation occurred under orbital stirring (100 rpm) at 28 °C for ten days. Some pellets were taken and weighed from pure cultures that were not contaminated.

The preparation of the semisolid fermentation (FSS) was carried out using fermentation trays containing 5 kg of pasteurized rice husk with a moisture content of 50%, to which 0.5 L of fresh pre-inoculum broth was added, using the prepared pellets with fungal growth. The incubation conditions were set at a temperature between 20 and 30 °C. The fermentation time frame was 20 days.

Samples (10 g) were taken every three days and leached with sterile water. Three replications were made for each sample. The samples were filtered under vacuum to separate the spent medium from the mycelium, which was subsequently weighed and frozen at -18 °C for subsequent lyophilization. Samples were stored and refrigerated in sterile bags until further chemical analysis.

2.3.2. Production of Enzymes

One hundred and fifty milliliters of medium was prepared with 10 g/L of glucose, 0.2 g/L of yeast extract, 0.5 g/L of ammonium tartrate, and 1 mL of Kirk's salts, at pH 5. The medium was sterilized for 15 min at 121 °C and inoculated with 5 mL of growth medium with each fungus. Each culture was maintained at 37 °C for 6 days, and the supernatant was analyzed for enzyme activity.

2.3.3. Analysis of Enzymatic Activity

In the culture supernatant, enzymatic activity assays were performed. The samples were analyzed in triplicate. A Shimadzu spectrophotometer (model UV-1800) was used to determine the following: (a) Lig. Peroxidase: This activity was determined by the Tien and Kirk Method. [3]. A 0.01 M of veratryl alcohol was used as a substrate in sodium tartrate buffer (0.1 M, pH 3.0). The reaction was initiated by adding 4 mM H_2O_2 and monitored by measuring the increase in absorbance at 310 nm that corresponds to veratraldehyde; (b) Methoxyl Hydrolase: The measurement method was applied on a methoxylated model compound. The reaction was followed by spectrophotometric measurements and the determination of the change in the spectrum. Two milliliters of 3,4,5-trimethoxybenzaldehyde 0.1 M, pH 4, was taken and incubated with 0.1 mL of the growth fungi supernatant at 30 °C. The decrease in the substrate was measured by UV spectrophotometry.

2.3.4. Production of Functional Compounds from Lignin

The production of the FC units was carried out in a semi-solid fermentation system for 20 days in a controlled chamber. The rice husk (25 kg) was inoculated with the propagated fungi Phanerochaete chrysosporium and Gloeophyllum trabeum previously tested for enzyme production capacity. The semi-solid fermentation process was performed in an open vessel without stirring. inside a controlled oven (humidity: 60–70%; temperature: 20–30 °C; pH 5–6). The activity of the inoculated fungi was controlled by measuring the enzymatic activity in the material of semi-solid fermentation. For this, 2 g of inoculated rice husk was taken and suspended in 10 mL of water. After 1 h, the enzymatic activity was determined in the liquid: lignin peroxidase at an average of 80 EU/mL and methoxyl hydrolase at an average of 50 UE/mL. Twenty days after enzymatic action, a volume equal to that occupied by the solid material was decanted for 2 h and centrifuged, and the supernatant was analyzed by UV spectrophotometry, which was also used to determine the units of FC (veratryl alcohol and veratryl aldehyde). The samples were analyzed in triplicate. The equipment used was a spectrophotometer of the Shimadzu brand (model

UV-1800). The FC extracts were lyophilized in a Biobase BK-FD10S vacuum freeze dryer and stored at 4 °C.

2.3.5. Enzymatic Process

From the analyzed samples, it was possible to determine the enzymatic activity in the extracts of both Lig peroxidase and methoxyl hydrolase enzymes. The extracts were lyophilized, and an enzymatic consortium was formulated in an aqueous medium with 0.2 M citrate buffer, pH 4.5, containing the following enzymatic constitution: 240 UE/mg Lig peroxidase and 150 UE/mg methoxy hydrolase. This enzymatic solution was the medium in which the residual lignocellulosic of the FSS was added in a proportion of 100 g of the solid substrate (40% moisture) in 200 mL of medium. This procedure was performed in shaker shaking at 30 °C for 2 h. After the oxidative depolymerization and hydrolysis of the methoxy groups by the acting enzymes, the material was filtered, and the supernatant liquid was stored at −18 °C for subsequent characterization analysis.

2.4. Characterization of the Obtained Product

2.4.1. Phenolic Compound Profile

The lyophilized extract was redissolved in ethanol/water (80:20, *v/v*), to determine the phenolic compounds profiles by chromatographic analysis using a Dionex Ultimate 3000 UPLC (Thermo Scientific, San Jose, CA, USA), following the protocol previously described by Bessada, Barreira, Barros, Ferreira, & Oliveira, (2016) [4]. Detection was carried out using a diode array detector (DAD) using the preferred wavelengths of 280 nm and 370 nm. Data acquisition was carried out using the Xcalibur® data system (Thermo Finnigan, San Jose, CA, USA). The identification was performed with the available standard compounds and by using literature information regarding the fragmentation pattern. Quantification was performed using seven-level calibration curves obtained from commercial standard compounds. The results were expressed in mg per g of extract.

2.4.2. Antioxidant Activity

Two cell-based assays were performed to evaluate the in vitro antioxidant activity of the samples. Trolox (Sigma-Aldrich, St. Louis, MO, USA) was the positive control used in both assays. The thiobarbituric acid reactive substances (TBARS) formation inhibition capacity of the extract was evaluated using porcine brain cell homogenates following the in vitro assay previously described by [5]. The results were expressed as EC50 values (µg/mL), i.e., extract concentration providing 50% of antioxidant activity. The antihemolytic activity of the extracts was evaluated using sheep erythrocytes by OxHLIA assay, as previously described by [6]. The results were given as IC50 values (µg/mL) for the Δt of 60 and 120 min, i.e., extract concentration required to keep 50% of the erythrocyte population intact for 60 and 120 min.

The cytotoxic activity was evaluated using four human tumor cell lines, namely MCF-7 (breast adenocarcinoma), NCI-H460 (non-small cell lung cancer), HeLa (cervical carcinoma), and HepG2 (hepatocellular carcinoma). All cell lines were commercially acquired from DSMZ (Leibniz-Institut DSMZ-Deutsche Sammlung von Mikroorganismen und Zellkulturen GmbH) and routinely maintained with RPMI-1690 medium enriched with 10% fetal bovine serum, L-glutamine (2 mM), penicillin (100 U/mL), and streptomycin (100 µg/mL). The cells were incubated under a humidified atmosphere at 37 °C and 5% CO_2, and were only used when they reached 80 to 90% confluence. A primary culture of non-tumor cells was also tested from a freshly harvested porcine liver (PLP2) and established according to the previously described [7]. The studied extracts were re-dissolved in water (8 mg/mL) and further diluted to obtain the range of concentrations to be tested (0.125–8 mg/mL). An aliquot (10 µL) of the extract's different concentrations was incubated for 48 h with the different tested cell lines (190 µL, 10,000 cells/mL). The cytotoxic potential was analyzed using the sulforhodamine B colorimetric assay and according to the procedure previously described [8]. Ellipticine was used as a positive control, and the results were expressed as extract concentration that causes 50% inhibition of cell proliferation (GI50 values).

2.4.3. Anti-Inflammatory Activity

The evaluation of the anti-inflammatory activity was performed determining the capacity of the extracts to inhibit the lipopolysaccharide (LPS)-induced nitric oxide (NO) production in a murine macrophage cell line (RAW 264.7). The extracts were re-dissolved and further diluted following the previously described (Section 2.4.2.). The procedure was performed according to what was previously described [9]. Dexamethasone corticosteroid (50 mM) was used as positive control, and cells with and without LPS were considered negative controls. The obtained results were expressed in IC_{50} values (µg/mL), corresponding to the concentration of the extracts responsible for 50% of NO production inhibition.

2.4.4. Antibacterial Activity

The antimicrobial activity was evaluated using six Gram-negative bacteria: Escherichia coli, Escherichia coli ESBL, Klebsiella pneumoniae, Klebsiella pneumoniae ESBL, Morganella morganii, and Pseudomonas aeruginosa; and four Gram-positive bacteria: Enterococcus faecalis, Listeria monocytogenes, MRSA (Methicillin resistant Staphylococcus aureus), and MSSA (methicillin susceptible Staphylococcus aureus). The clinical isolates were obtained from patients hospitalized in various departments at the North-eastern local health unit (Bragança, Portugal) and Hospital Center of Trás-os-Montes and Alto Douro (Vila Real, Portugal). To maintain the exponential growth phase, all microorganisms were incubated at 37 °C in an appropriate fresh medium for 24 h before analysis. The extracts were redissolved in water (100 mg/mL) and further diluted to obtain a range of seven concentrations below the stock solution to determine the antibacterial activity. The minimal inhibitory concentration (MIC) determination was conducted using a colorimetric assay using p-iodonitrotetrazolium chloride according to a previously described procedure [10]. Two negative controls were prepared, one with Mueller-Hinton Broth (MHB). Two positive controls were prepared with MHB; each contained inoculum, but one contained the medium, antibiotic, and bacteria for Gram-negative bacteria. Ampicillin and imipenem were used as positive controls, while ampicillin and vancomycin were used for Gram-positive bacteria.

To determine the MBC (Minimum Bactericidal Concentration), 10 µL of liquid from each well that showed no change in color was plated on a solid medium, Blood agar (7% sheep blood), and incubated at 37 °C for 24 h. The lowest concentration that yielded no growth was the MBC. The MBC is defined as the lowest concentration required to kill bacteria.

2.5. Statistical Analysis

The experiments were carried out in triplicate, and the results were expressed as mean ± standard deviation. SPSS Statistics software (IBM SPSS Statistics for Windows, Version 22.0. Armonk, NY, USA: IBM Corp.) was used to assess significant differences ($p < 0.05$) among samples by applying a two-tailed paired Student's t-test at a 5% significance level.

3. Results and Discussion

3.1. Protein and Phenolic Compounds Yield

Protein concentrations of 1.2 mg/mL was obtained in the filtrates of enzyme production by mixed cultures of *G. trabeum* and *P. chrysosporium*. The enzymatic activities obtained are shown in Table 1. Enzymatic activity of lignin peroxidase and methoxyl hydrolase was determined, being higher for lignin peroxidase.

Table 1. Enzymatic activity obtained in the liquid system of *Gloeophyllum trabeum* and *Phanerochaete chrysosporium*.

Enzyme	*G. trabeum* (UE/mL)	*P. chrysosporium* (UE/mg)
Lignin peroxidase	80	50
Methoxyl hydrolase	50	31

Solid fermentation (SF) was carried out on 500 g of rice husk, with an enzymatic extract of 80 UE/mL, and 50 UE/mL with a specific activity of 50 y 30 UE/mg of proteins. Particularly, 50 g of rice husk in SSF was lixiviated with 200 mL of water, and 200 mL of aqueous extract equivalent to 6.5 g FC (determined after the lyophilization) was obtained. The performance of the process reveals that from 100 g of lignocellulosic material 56 g of material at the end of the transformation was obtained. The other 44 g were solubilized and depolymerized to obtain the phenols extract, 12% of the rice husk, and 85% of the rice husk lignin content (14%). The products of the rice husk treatment were lyophilized and characterized to identify and quantify the compounds present in the mixture (mg/g extract).

The biotransformation yields have been very interesting from the point of view of the % of the oligomers solubilized in lignin since 45% of oligomers and soluble monomers were obtained at pH 7. This fraction represents the modification obtained in the combined solid fermentation process followed by an enzymatic transformation that allows for shorter treatment times. In solid fermentation processes, only 40% can be obtained in 60 days, while the time could be shortened to 20 days in this new stage procedure.

3.2. Phenolic Compounds Profile

Two organic compounds were found in the studied samples of rice husk (Table 2), which were identified by comparing their retention time and maximum spectra with two available standard compounds, namely veratryl alcohol (peak 1) and veratryl aldehyde (peak 2), that have identified in rice husk samples [11–13].

Table 2. Retention time (Rt), wavelengths of maximum absorption in the visible region (λmax), tentative identification, and quantification (mg/g extract) of the organic compounds present in rice husk (mean ± SD).

Peak	Rt (min)	λmax (nm)	Identification	Quantification (mg/g Extract)
1	11.46	276	Veratryl alcohol	70.4 ± 0.1
2	21.77	278/sh309	Veratryl aldehyde	23.3 ± 0.3
-	-	-	Total	93.7 ± 0.2

Standard calibration curves: veratryl alcohol ($y = 51{,}266x + 414{,}240$, $R^2 = 0.999$) and veratryl aldehyde ($y = 43{,}916x + 305{,}634$, $R^2 = 0.999$).

3.3. Bioactive Properties

The antioxidant activity was measured in vitro using two cell-based bioassays, namely TBARS and OxHLIA, which assess the extracts' ability to inhibit the formation of thiobarbituric acid reactive substances (TBARS), such as malondialdehyde (MDA), which results from the degradation of lipid peroxidation products (TBARS assay), and to protect the erythrocyte membranes from oxidative hemolysis initially induced by 2,2'-azobis(2-methylpropionamidine) dihydrochloride (AAPH)-derived free radicals (OxHLIA assay), respectively. The results obtained with these assays are shown in Table 3, where the extract concentration required to provide 50% of antioxidant activity via TBARS formation inhibition or to keep 50% of the erythrocyte population intact for 60 and 120 min is presented. Thus, the lower the extract concentration (EC_{50} or IC_{50}), the greater its antioxidant activity. Furthermore, when measuring the antioxidant activity of natural extracts, some antioxidants may react more quickly and become depleted in the system, while others may offer prolonged antioxidant protection. The OxHLIA bioassay thus allowed for a distinction to be made between short-term (60 min Δt) and long-term (120 min Δt) antioxidant protection.

The rice husk extract yielded an EC_{50} value of 804 µg/mL in the TBARS bioassay, and IC_{50} values of 136 and 341 µg/mL in OxHLIA for Δt of 60 and 120 min, respectively (Table 3). These values are higher than Trolox's, but while this synthetic analog of α-tocopherol is a pure antioxidant, the rice husk extract is a complex mixture of various compounds, including non-active constituents. In general, it appeared that the extract performed better in the OxHLIA bioassay than in the TBARS, as the result was closer to the positive control.

Table 3. Antioxidant, anti-inflammatory, and cytotoxic activities of rice husk extract and positive control (mean ± SD).

		Rice Husk	Positive Control
Antioxidant activity			Trolox
TBARS inhibition (EC_{50}, µg/mL)		804 ± 39	5.8 ± 0.6
OxHLIA (IC_{50}, µg/mL)	60 min Δt	136 ± 5	19 ± 2
	120 min Δt	341 ± 17	41 ± 2
Anti-inflammatory activity			Dexamethasone
NO production inhibition (EC_{50}, µg/mL)		>400	16 ± 1
Cytotoxicity (GI_{50}, µg/mL)			Ellipticine
MCF-7 (breast carcinoma)		310 ± 6	0.91 ± 0.04
NCI-H460 (non-small cell lung carcinoma)		>400	1.03 ± 0.09
HeLa (cervical carcinoma)		>400	1.91 ± 0.06
HepG2 (hepatocellular carcinoma)		239 ± 3	1.1 ± 0.2
PLP2 (non-tumor porcine liver primary cells)		>400	3.2 ± 0.7
Anti-inflammatory activity (IC_{50} values, µg/mL)			
RAW 264.7 (murine macrophage)		>400	16 ± 1

Statistical differences ($p < 0.05$) between extract and positive control were found using a two-tailed paired Student's t-test.

The antiproliferative capacity of rice husk extract was tested against several cell lines. The obtained results are presented in Table 3 as the extract concentration required to inhibit the cell proliferation in 50% (GI_{50} values). The selection of the tested tumor cell lines was based on the higher incidence associated with those types of cancer. The studied extract only demonstrates the capacity to inhibit the proliferation of MCF-7 and HepG2 cell lines. The highest susceptibility was observed for the HepG2 cell line, exhibiting the lowest GI_{50} values (GI_{50} = 293 µg/mL). No activity was observed for the remaining tumor cell lines tested (NCI-H460, HeLa) as they showed GI_{50} values higher than the highest concentration tested (GI_{50} > 400 µg/mL). No cytotoxicity was observed regarding the primary culture of non-tumor cells (GI_{50} > 400 µg/mL), highlighting the security associated with the studied agro-waste material.

Similar results were described by Gao and co-workers [11], having demonstrated the antiproliferative potential of rice musk for the HepG2 cell line. The cytotoxic potential of this agro-waste material is scarcely studied. Most of those studies evaluated the cytotoxic properties of peptides obtained from rice husk. The highest cytotoxic power was described for samples with the highest hydrolyzed protein content. The authors considered that the presence of glutamic acid and proline are the main ones responsible for the reported activity [12]. Further studies are needed to understand the compounds responsible for the demonstrated potential and the mechanisms involved.

The anti-inflammatory potential of rice husk extract was assessed through the NO inhibition production by the LPS-stimulated murine macrophage cells (RAW 246.7). The extract concentration with the capacity to inhibit in 50% de NO production (IC_{50} values) is exhibited in Table 3. No activity was observed with the cell-based assay used at the tested concentrations (between 400 and 6.25 µg/mL). The extract does not exhibit a capacity to inhibit the NO production with GI_{50} values >400 µg/mL.

The phenolic compounds identified in rice samples, namely ferulic acid and quercetin, are described as potent anti-inflammatories [13]. Also, studies with rice bran describe their components as having anti-inflammatory potential [14]. Although the activity has not been demonstrated, its study is very scarce, and it would be interesting to use assays that evaluate other mechanisms involved in the inflammatory process.

The antioxidant activity of rice husk subjected to solid-state fermentation with different strains of *Pleurotus sapidus* was previously evaluated by Pinela et al. (2020) [15] using the same bioassays. With OxHLIA, the authors reached higher IC_{50} values (179 and 259 µg/mL for 60 min Δt when using monokaryotic and dikaryotic strains, respectively) than in the present work. Curiously, the residual or spent mushroom substrate obtained after the fructification of the dikaryotic strain had a lower IC_{50} value (83 µg/mL). Still, the antihemolytic activity of unfermented rice husk (IC_{50} of 42.8 µg/mL for 60 min Δt) was higher than that of fermented samples. For TBARS, both unfermented and monokaryotic strain-fermented samples presented the same result (~155 µg/mL), which are much higher what was observed in the present work (Table 1) in the spent mushroom substrate (317 µg/mL). In general, the authors concluded that the fermentation process decreased the antioxidant activity, and the content of phenolic compounds (mainly phenolic acids).

Regarding the results obtained for antibacterial activity (Table 4), the extract of rice husk presented the capacity to inhibit the growth of all tested bacteria in a range of 20 to 5 mg/mL. Methicillin-resistant *Staphylococcus aureus* was the microorganisms with better antimicrobial results in a MIC of 5 mg/mL. These results were expected because *Staphylococcus aureus* is a Gram-positive bacteria and was more sensitive than Gram-negative bacteria due to differences in the chemical and physical properties of the cell wall. Moreover, the presence of sugar parts of lignin can interact with the peptidoglycan of the cell membrane of bacteria. The results obtained are in accordance with Tran et al., 2021 [16], who focused their study on the sustainably rice-husk-extracted lignin, nano-lignin (n-Lignin), lignin-capped silver nanoparticles (LCSNs), n-Lignin-capped silver nanoparticles (n-LCSNs), and lignin-capped silica-silver nanoparticles (LCSSNs), using them for antibacterial activities. The results showed that the antimicrobial activity of all compounds was better against Gram-positive bacteria, *S. aureus*, than against Gram-negative bacteria, *E. coli*.

Table 4. Antibacterial activity (MIC and MBC values mg/mL) of rice husk (mean ± SD).

	Rice Husk		Ampicillin		Imipenem		Vancomycin	
	MIC	MBC	MIC	MBC	MIC	MBC	MIC	MBC
Gram-negative bacteria								
Escherichia coli	10	>20	<0.15	<0.15	<0.0078	<0.0078	n.t.	n.t.
Klebsiella pneumoniae	10	>20	10	20	<0.0078	<0.0078	n.t.	n.t.
Morganella morganii	10	>20	20	>20	<0.0078	<0.0078	n.t.	n.t.
Proteus mirabilis	20	>20	<015	<0.15	<0.0078	<0.0078	n.t.	n.t.
Pseudomonas aeruginosa	10	>20	>20	>20	0.5	1	n.t.	n.t.
Gram-positive bacteria								
Enterococcus faecalis	10	>20	<0.15	<0.15	n.t.	n.t.	<0.0078	<0.0078
Listeria monocytogenes	10	>20	<0.15	<0.15	<0.0078	<0.0078	n.t.	n.t.
MRSA	5	>20	<0.15	<0.15	n.t.	n.t.	0.25	0.5

MIC and MBC correspond to the minimal sample concentration inhibiting the bacterial growth or killing the original inoculum. n.t.—Not tested.

The tested sample presented a bacteriostatic effect but not bactericidal.

4. Conclusions

From the tests carried out, it can be concluded that using an FSS process for rice residues can be an important management and recovery solution for rice husk that can provide circularity to the processes, obtaining products of great added value for bioremediation. On the other hand, the extraction of the enzymes produced by the microorganisms allows for their use in a second liquid process where compounds such as veratraldehyde and veratrylic alcohol are obtained with bacteriostatic properties with an interesting activity for Gram-negative bacteria, such as *Escherichia coli*, *Klebsiella pneumonia*, *Pseudomonas aeruginosa*, *Proteus mirabilis*, and *Morganella morganii*, and Gram-positive bacteria, such as *Enterococcus faecalis* and *Listeria monocytogenes*. The use of veratraldehyde alcohol and veratrylic aldehyde in the obtained extracts marks an interesting application for materials to be functionalized with germicidal properties, such as food containers and clothing and sanitary accessories. In future works, the encapsulation of the extracts for their controlled release and incorporation into different matrices for different applications will be studied.

Author Contributions: M.I.D.: Methodology; investigation; writing—original draft preparation; F.M.: Methodology.; investigation; writing—original draft preparation. T.C.S.P.P.: Methodology; investigation; writing—original draft preparation. J.P.: Methodology; investigation; writing—original draft preparation. M.-F.B.: Methodology; investigation; writing—original draft preparation; writing—review and editing. L.B.: Methodology; investigation; resources; writing—original draft preparation. J.R.V.-B.: Writing—review and editing; supervision. I.C.F.R.F.: Conceptualization; methodology. M.L.: Conceptualization; methodology; investigation; resources; writing—original draft preparation.; writing—review and editing; supervision. All authors have read and agreed to the published version of the manuscript.

Funding: Thanks to FCT for financial support through national funds FCT/MCTES, to CIMO (UIDB/00690/2020 and UIDP/00690/2020), and to SusTEC (LA/P/0007/2021). Centro de Investigaciones Nucleares, Facultad de Ciencias, Universidad de la República, Uruguay, through their own funds, with no other contribution. RED CYTED ENVABIO100 (Ref: 121RT0108) (interaction and publication cost).

Institutional Review Board Statement: Not applicable.

Informed Consent Statement: Not applicable.

Data Availability Statement: Data sharing is not applicable to this article.

Acknowledgments: RED CYTED BIORRECER and ENVABIO100 (Ref: 121RT0108).

Conflicts of Interest: The authors declare no conflict of interest.

References

1. Mattila, P.; Suonpää, K.; Piironen, V. Functional properties of edible mushrooms. *Nutrition* **2000**, *16*, 694–696. [CrossRef] [PubMed]
2. Çağlarırmak, N. The nutrients of exotic mushrooms (*Lentinula edodes* and *Pleurotus* species) and an estimated approach to the volatile compounds. *Food Chem.* **2007**, *105*, 1188–1194. [CrossRef]
3. Potumarthi, R.; Baadhe, R.R.; Nayak, P.; Jetty, A. Simultaneous pretreatment and sacchariffication of rice husk by *Phanerochete chrysosporium* for improved production of reducing sugars. *Bioresour. Technol.* **2013**, *128*, 113–117. [CrossRef] [PubMed]
4. Bessada, S.M.F.; Barreira, J.C.M.; Barros, L.; Ferreira, I.C.F.R.; Oliveira, M.B.P.P. Phenolic profile and antioxidant activity of Coleostephus myconis (L.) Rchb.f.: An underexploited and highly disseminated species. *Ind. Crops Prod.* **2016**, *89*, 45–51. [CrossRef]
5. Pinela, J.; Barros, L.; Carvalho, A.M.; Ferreira, I.C.F.R. Nutritional composition and antioxidant activity of four tomato (*Lycopersicon esculentum* L.) farmer' varieties in Northeastern Portugal homegardens. *Food Chem. Toxicol.* **2012**, *50*, 829–834. [CrossRef] [PubMed]
6. Lockowandt, L.; Pinela, J.; Roriz, C.L.; Pereira, C.; Abreu, R.M.V.; Calhelha, R.C.; Alves, M.J.; Barros, L.; Bredol, M.; Ferreira, I.C.F.R. Chemical features and bioactivities of cornflower (*Centaurea cyanus* L.) capitula: The blue flowers and the unexplored non-edible part. *Ind. Crops Prod.* **2019**, *128*, 496–503. [CrossRef]
7. Abreu, R.M.V.; Ferreira, I.C.F.R.; Calhelha, R.C.; Lima, R.T.; Vasconcelos, M.H.; Adega, F.; Chaves, R.; Queiroz, M.-J.R.P. Anti-hepatocellular carcinoma activity using human HepG2 cells and hepatotoxicity of 6-substituted methyl 3-aminothieno [3,2-b]pyridine-2-carboxylate derivatives: In vitro evaluation, cell cycle analysis and QSAR studies. *Eur. J. Med. Chem.* **2011**, *46*, 5800–5806. [CrossRef] [PubMed]
8. Barros, L.; Pereira, E.; Calhelha, R.C.; Dueñas, M.; Carvalho, A.M.; Santos-Buelga, C.; Ferreira, I.C.F.R. Bioactivity and chemical characterization in hydrophilic and lipophilic compounds of *Chenopodium ambrosioides* L. *J. Funct. Foods* **2013**, *5*, 1732–1740. [CrossRef]
9. Souilem, F.; Fernandes, Â.; Calhelha, R.C.; Barreira, J.C.M.; Barros, L.; Skhiri, F.; Ferreira, I.C.F.R. Wild mushrooms and their mycelia as sources of bioactive compounds: Antioxidant, anti-inflammatory and cytotoxic properties. *Food Chem.* **2017**, *230*, 40–48. [CrossRef] [PubMed]
10. Svobodova, B.; Barros, L.; Calhelha, R.C.; Heleno, S.; Alves, M.J.; Walcott, S.; Bittova, M.; Kuban, V.; Ferreira, I.C.F.R. Bioactive properties and phenolic profile of *Momordica charantia* L. medicinal plant growing wild in Trinidad and Tobago. *Ind. Crops Prod.* **2017**, *95*, 365–373. [CrossRef]

11. Gao, Y.; Guo, X.; Liu, Y.; Mingwei, Z.; Ruifen, Z.; Lijun, Y.; Tong Rui, H.L. A full utilization of rice husk to evaluate phytochemical bioactivities and prepare cellulose nanocrystals. *Sci. Rep.* **2018**, *8*, 10482. [CrossRef] [PubMed]
12. Ilhan-Ayisigi, E.; Budak, G.; Celiktas, M.S.; Sevimli-Gur, C.; Yesil-Celiktas, O. Anticancer activities of bioactive peptides derived from rice husk both in free and encapsulated form in chitosan. *J. Ind. Eng. Chem.* **2021**, *103*, 381–391. [CrossRef]
13. Wisetkomolmat, J.; Arjin, C.; Hongsibsong, S.; Ruksiriwanich, W.; Niwat, C.; Tiyayon, P.; Jamjod, S.; Yamuangmorn, S.; Prom-U-Thai, C.; Sringarm, K. Antioxidant Activities and Characterization of Polyphenols from Selected Northern Thai Rice Husks: Relation with Seed Attributes. *Rice Sci.* **2023**, *30*, 148–159. [CrossRef]
14. Lopretti, M.I.; Lecot Calandria, N.V.; Rodriguez, A.; Lluberas Nuñez, M.G.; Orozco, F.; Bolaños, L.; Vega-Baudrit, J. Biorefinery of rice husk to obtain functionalized bioactive compounds. *J. Renew. Mater.* **2019**, *7*, 313–324. [CrossRef]
15. Pinela, J.; Omarini, A.B.; Stojković, D.; Barros, L.; Postemsky, P.D.; Calhelha, R.C.; Breccia, J.; Fernández-Lahore, M.; Soković, M.; Ferreira, I.C.F.R. Biotransformation of rice and sunflower side-streams by dikaryotic and monokaryotic strains of Pleurotus sapidus: Impact on phenolic profiles and bioactive properties. *Food Res. Int.* **2020**, *132*, 109094. [CrossRef] [PubMed]
16. Tran, N.T.; Trang, T.T.N.; Ha, D.; Nguyen, T.H.; Nguyen, N.N.; Baek, K.; Nguyen, N.T.; Tran, C.K.; Tran, T.T.V.; Le, H.V.; et al. Highly Functional Materials Based on Nano-Lignin, Lignin, and Lignin/Silica Hybrid Capped Silver Nanoparticles with Antibacterial Activities. *Biomacromolecules* **2021**, *22*, 5327–5338. [CrossRef] [PubMed]

Disclaimer/Publisher's Note: The statements, opinions and data contained in all publications are solely those of the individual author(s) and contributor(s) and not of MDPI and/or the editor(s). MDPI and/or the editor(s) disclaim responsibility for any injury to people or property resulting from any ideas, methods, instructions or products referred to in the content.

Proceeding Paper

Lactide Synthesis Using ZnO Aqueous Nanoparticles as Catalysts †

Shirley Duarte [1,*], Axel Dullak [1], Francisco P. Ferreira [2], Marcelo Oddone [1,*] and Darío Riveros [1]

1. Facultad de Ciencias Químicas, Universidad Nacional de Asunción, San Lorenzo 1055, Paraguay; adullak@qui.una.py (A.D.); darioriverosromero@gmail.com (D.R.)
2. Laboratory of Organic Chemistry and Natural Products-LAREV, Department of Biology, Faculty of Exact and Natural Sciences, National University of Asunción, San Lorenzo 3291, Paraguay; franciscoferreira@facen.una.py
* Correspondence: sduarte@qui.una.py (S.D.); marceo015@gmail.com (M.O.)
† Presented at the 1st International Conference of the Red CYTED ENVABIO100 "Obtaining 100% Natural Biodegradable Films for the Food Industry", San Lorenzo, Paraguay, 14–16 November 2022.

Abstract: The increasing global consumption of conventional plastics has led to significant environmental challenges due to their resistance to degradation and dependency on petroleum, a volatile resource. In this context, polylactic acid (PLA) has emerged as a promising biopolymer alternative, offering excellent physical and mechanical properties while being derived from renewable resources. The synthesis of PLA involves the production of lactide, a crucial cyclic monomer. This study focuses on lactide synthesis using zinc oxide (ZnO) nanoparticles as catalysts. The synthesis process consists of three stages: the dehydration of lactic acid, oligomerization to obtain a PLA oligomer, and depolymerization with simultaneous distillation to produce lactide. The ZnO nanoparticle catalyst proved to be highly efficient in regulating the molecular weight of the oligomer during depolymerization, which directly impacts the molecular weight of the PLA. The lactide purification using ethyl acetate resulted in an average purity of 90.0 ± 1.79%, demonstrating the effectiveness of the purification process. The quantification of lactide through high-performance liquid chromatography (HPLC) showed excellent linearities, allowing for the accurate determination of lactide content. The lactide synthesis yielded 77–80%, and the stability of the synthesized lactide was confirmed through a second purity determination after four months, with only a 1.7% loss in lactide content. Overall, this study showcases the feasibility of lactide synthesis using ZnO nanoparticles as catalysts and contributes valuable insights into producing high-quality lactides for PLA manufacturing.

Keywords: lactide synthesis; zinc oxide nanoparticle; polylactic acid

Citation: Duarte, S.; Dullak, A.; Ferreira, F.P.; Oddone, M.; Riveros, D. Lactide Synthesis Using ZnO Aqueous Nanoparticles as Catalysts. *Biol. Life Sci. Forum* **2023**, *28*, 13. https://doi.org/10.3390/blsf2023028013

Academic Editor: José Vega-Baudrit

Published: 9 May 2023

Copyright: © 2023 by the authors. Licensee MDPI, Basel, Switzerland. This article is an open access article distributed under the terms and conditions of the Creative Commons Attribution (CC BY) license (https://creativecommons.org/licenses/by/4.0/).

1. Introduction

The global consumption of conventional plastics has witnessed a steady rise alongside improvements in quality of life. Unfortunately, these plastics pose significant environmental challenges, accumulating in landfills and oceans worldwide due to their resistance to environmental and biological degradation [1]. Furthermore, concerns over the volatile costs of petroleum, the primary precursor of conventional plastics, have added to the urgency of finding sustainable alternatives [2]. As a result, extensive research efforts have been directed towards exploring renewable and biodegradable sources of plastics [1].

Among the various alternatives, polylactic acid (PLA) has emerged as a promising biopolymer, garnering significant interest in recent years. PLA exhibits excellent physical and mechanical properties and can be processed using existing machinery with only minor adjustments [3].

Using lactide as a biopolymer offers significant environmental benefits due to its renewable and biodegradable nature. Unlike conventional plastics derived from petroleum, PLA is produced from renewable resources, which reduces dependency on fossil fuels and mitigates the environmental impact of plastic production. Additionally, PLA exhibits superior biodegradability, breaking down into harmless natural compounds over time, thus minimizing plastic waste accumulation and its adverse effects on ecosystems.

The most widely employed method for industrial PLA production is ring-opening polymerization (ROP), a process that involves the propagation of cyclic monomers initiated by various ions [4] and applied to olefins. Lactide, the intermediate cyclic monomer in the PLA production process, holds particular significance [5,6].

The synthesis of lactide typically begins with the distillation of lactic acid, followed by its oligomerization, wherein heating lactic acid at high temperatures leads to the release of condensed water. Subsequently, some of the resulting oligomers with specific molecular weights undergo depolymerization to yield lactide [7–9].

However, lactide synthesis via ROP is known for its high cost and low yield [9]. Therefore, researchers have focused on exploring the most suitable catalytic systems and optimal reaction conditions to enhance the overall yield and properties of synthesized PLA [4,6].

In this context, homogeneous tin-based catalysts have been widely studied for their role in increasing the molecular weight of the oligomer, thus impeding its depolymerization to lactide. Conversely, ZnO nanoparticle catalysts have shown promise in this stage of the process [10,11]. ZnO effectively maintains the depolymerization equilibrium necessary for lactide production by regulating the molecular weight of the oligomer [10]. The molecular weight of the oligomer significantly impacts the molecular weight of the resulting PLA, as reduced mobility affects the occurrence of the "back-biting" reaction [12].

In this work, we present the synthesis of lactide through ROP catalyzed with ZnO nanoparticles, starting from a solution of commercial lactic acid. We assess the efficiency of this process by determining the production and conversion yield while also measuring the purity of the lactide using HPLC and its thermal properties through DSC-TGA analysis. Herein, we provide a comprehensive account of our observations and discuss the obtained results.

2. Materials and Methods

2.1. Materials

L-lactic acid was procured from Mater Food, Paraguay, with a monomer concentration of 88.2% in water. ZnO nanoparticles were obtained from US Research Nano-materials Inc., Houston, TX, USA, as an aqueous dispersion of Nano-ZnO, 30–40 nm, with a concentration of 20 wt% in water. Ethyl acetate (analytical-grade organic solvent) was used for lactide purification. Chromatographic-grade acetonitrile and water (Merck) were utilized for lactide quantification. All chemicals were used directly without further purification.

2.2. Lactide Synthesis

Lactide synthesis was carried out using ZnO nanoparticle dispersion (30–40 nm, 20 wt%) as a highly efficient catalyst at a load ratio of 0.6 wt%, following the method reported by Hu et al. [10].

The process consists of three stages, each with specific conditions: the dehydration of lactic acid, oligomerization to obtain a PLA oligomer, and depolymerization with simultaneous lactide distillation. The process was performed in triplicate.

The lactide synthesis was conducted using a laboratory-scale system, as depicted in Figure 1.

Initially, an aqueous solution of the commercial lactic acid (250 g) and Nano-ZnO catalyst (20 wt%) at a load ratio of 0.6 wt% was introduced into a round-bottom flask. The temperature was raised to 80 °C using a thermal oil bath, and a vacuum pump decreased the pressure down to 60 kPa, as optimized by Hu et al. [10]. Dehydration proceeded until no more water was expelled from the solution.

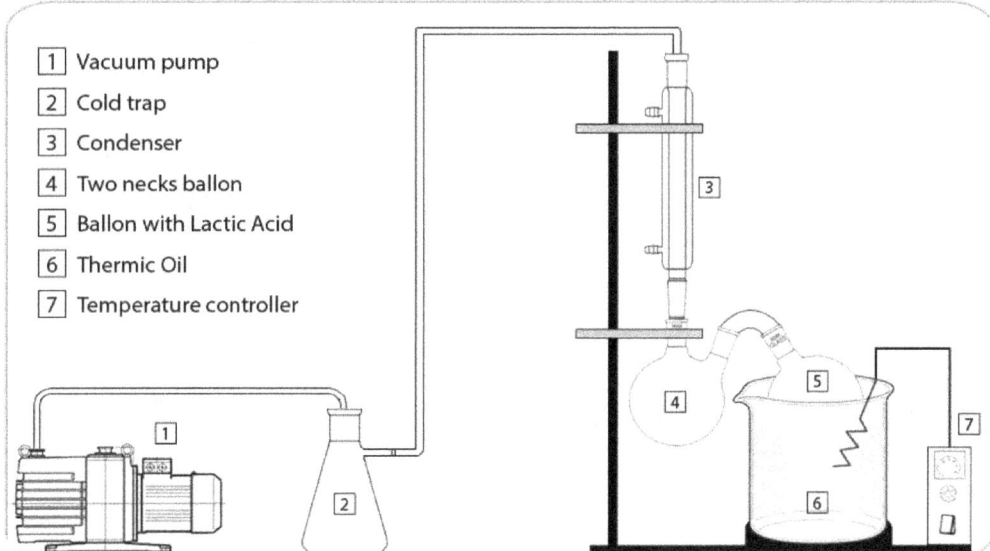

Figure 1. Scheme for the dehydration process.

For the oligomerization of lactic acid, it was important to avoid excessively high molecular weight, as this would hinder the subsequent formation of lactide. Therefore, the temperature was gradually increased to 150 °C, while the pressure was reduced to 10 kPa. During this step, water was produced by the formation of ester bonds between the hydroxyl and carboxyl groups and had to be removed [7]. To achieve this, the conditions were maintained for three hours or until no further water condensed in the double-neck flask. The product obtained at this stage was the PLA oligomer, characterized by a lower molecular weight due to the fewer repeating units in the chain [13].

After completing the oligomerization, the subsequent stage was undertaken, using the same setup, but without the condenser. To depolymerize the oligomer and obtain lactide, the pressure was gradually reduced to 3 kPa, and the temperature was raised to 220 °C. This allowed the removal of lactide from the reaction system through distillation [10]. Lactide was collected in a round-bottom flask submerged in an ice bath for rapid solidification. The process continued until no more product was obtained from the heated flask.

2.3. Lactide Purification

To ensure the suitability of lactide for PLA production, it is essential to remove any residual acid and water. Recrystallization in an appropriate solvent is an effective purification method, with ethyl acetate demonstrating excellent capability for this purpose [10,14].

The recrystallization process followed the conditions reported by Hu et al. [10] and Sanglard et al. [14]. The crude lactide was dissolved in ethyl acetate (1:1.5 w/v) and stirred at 80–90 °C for 10 min. The hot solution was filtered to remove any undissolved impurities and the filtrate was then cooled to room temperature and subsequently to 4 °C for lactide recrystallization. The resulting crystals were collected using vacuum suction and dried in a vacuum oven at 40 °C to complete the purification process.

To account for any potential lactide loss during recrystallization, three batches of purified lactide were re-purified together in ethyl acetate. This approach ensured a unified lactide purity level and facilitated a comparison of yields for each recrystallization process. The recrystallization yield was calculated using Equation (1).

$$\eta_{\text{recry}} = \frac{\text{Crude lactide [g]}}{\text{Purified lactide [g]}} \times 100 \quad (1)$$

2.4. Lactide Quantification

The quantity of lactide produced was determined through dry-weight measurement, accounting for the conversion yield of lactic acid into lactide (Equation (2)) and the production yield (Equation (3)). The theoretical production of lactide was calculated based on the actual lactic acid content in the solution.

$$\eta_{\text{conv}} = \frac{\text{Lactide [g]}}{\text{Lactic acid [g]}} \times 100 \quad (2)$$

$$\eta_{\text{prod}} = \frac{\text{Actual lactide production [g]}}{\text{Theoretical lactide production [g]}} \times 100 \quad (3)$$

2.5. Lactide Characterization

For the determination of purity, a quantitative HPLC method was employed, analyzing the lactide content using a standard calibration curve prepared in the range of 25–125 µg/mL.

The quantification was performed using Shimadzu LC-20A HPLC equipment, with a low-pressure quaternary pump, an autosampler in which 25 µL of the sample is injected, and an Octadecylsilane (C18) column measuring 25 cm × 4 mm internal diameter × 5 microns of padding.

The mobile phase consisted of a 2:8 (v/v) mixture of acetonitrile and water, with a flow rate of 0.8 mL/min. The detector wavelength was set at 250 nm, and the oven temperature was maintained at 30 °C. Lactide samples (0.1 g) were dissolved in anhydrous alcohol in 25 mL flasks, followed by shaking and filtration. The resulting filtrate was used for HPLC analysis.

The quantification method was based on the work of Feng et al. [15], who developed an innovative method for lactide quantification using hydrolytic kinetics. To assess the stability of the synthesized lactide, a second determination was performed.

Thermal analysis of the synthesized lactide was conducted using differential scanning calorimetry (DSC) and thermogravimetric analysis (TGA). A sample of the synthesized lactide (5 mg) was analyzed using a NETZSCH STA 449 F3 Jupiter simultaneous thermal analyzer (Selb, Germany), with a heating rate of 15 °C/min from 25 °C to 410 °C under a nitrogen atmosphere with a flow rate of 50 mL/min.

3. Results

3.1. Lactide Purification

Lactide synthesis was performed in triplicate, and Figure 2 shows the appearance of the crude synthesized lactide, while Figure 3 displays the recrystallized lactide after purification.

From the figure, we can see that the purification of lactide produces the formation of white and homogeneous crystals.

Figure 2. Crude synthesized lactide.

Figure 3. Recrystallized lactide.

3.2. Lactide Quantification

The results of the lactide synthesis and subsequent purification yields are summarized in Table 1.

Table 1. Yields in the synthesis and purification of lactide.

Run Number	η_{conv}	η_{prod}	η_{recry}^1	η_{recry}^2
1	68.71	77.91	54.51	
2	67.75	76.70	47.63	36.59
3	71.04	79.55	57.40	

The maximum conversion of concentrated lactic acid to lactide was 71.04%, which is in agreement with the value reported by Hu et al., where 3 more grams of lactide was obtained for every 100 g of lactic acid [10].

Regarding crystallization, a maximum yield of 57.4% was reached.

3.3. Lactide Characterization

The quantification of lactide using HPLC exhibited excellent linearity, with an R^2 value of 0.9998 in the equation of the line area = 5851.3[X] − 33,532, as shown in Figure 4, obtained from the injection of concentrated lactic acid standards between 25 and 125 µg mL^{-1}. The purity of the synthesized lactide, determined through triplicate analysis of the sample dilutions, was found to be an average of 90.0% ± 1.79 g.

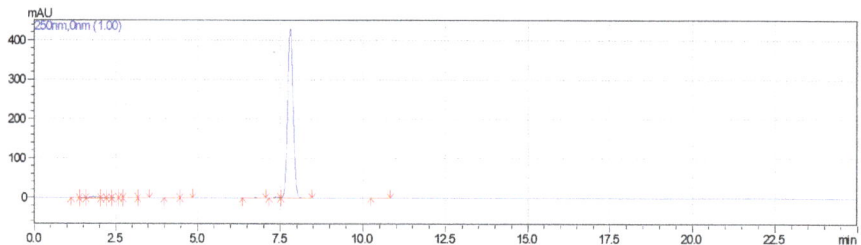

Figure 4. Chromatogram of standard solution (50 µg/mL).

The synthesized lactide was stored in a vacuum-sealed desiccator to maintain its stability. Determination and subsequent applications were performed promptly to minimize potential degradation. The stability of the synthesized lactide was evaluated by repeating the purity determination 120 days later, resulting in an 88.3% purity.

Thermogravimetric and differential scanning calorimetry analyses (TGA/DSC) (Figures 5 and 6) show the melting point (T_m), decomposition temperature (T_d), and enthalpy of the fusion (ΔH_m) of the lactide.

The mass decomposition curve and the respective derivative are presented in Figure 5, revealing $T_d = 227.8$ °C. Decomposition initiated at 108.32 °C with a weight loss of 0.54%, and it was nearly complete at 237.11 °C, with a weight loss exceeding 99.9%.

The melting properties of the synthesized lactide are depicted in Figure 6, with $T_m = 97.2$ °C and $\Delta H_m = 98.05$ J/g.

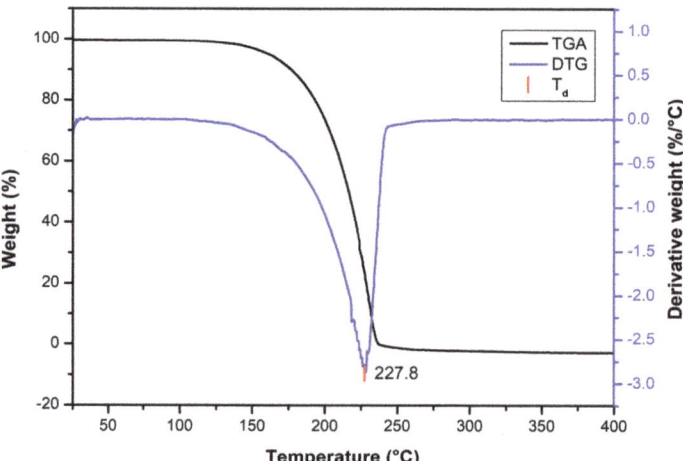

Figure 5. Thermogravimetric analysis of synthesized lactide.

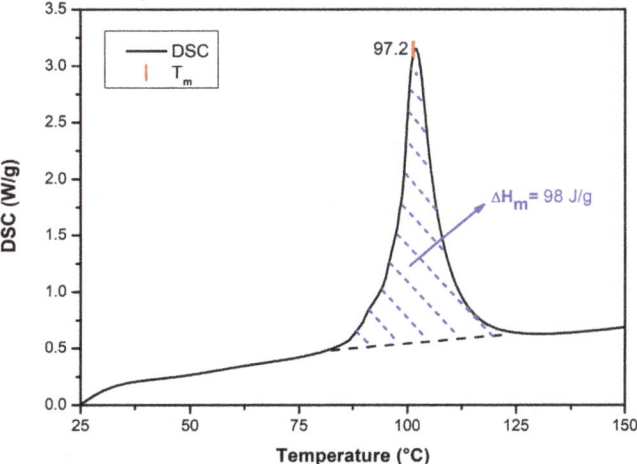

Figure 6. Differential scanning calorimetry analysis of synthesized lactide.

4. Discussion

4.1. Conversion and Production Yields of Lactide

The production yield of lactide varies significantly among different works, depending on the chosen method, catalyst, and reaction conditions [16].

In this study, we adopted the conditions determined by Hu et al. [10], who achieved a remarkably higher production yield of 91%, coupled with a high lactide purity of 95%. This notable result might be attributed to the low pressure (1 kPa) used in their process, which not only prompted the reaction to progress forward but also facilitated the removal of water, a crucial factor for lactide molecule formation [9,17].

In our experimental setup, the vacuum pump could only reach a working pressure of 3 kPa, whereas other investigations employing the distillation of lactide operated at much lower pressures, typically between 1300 and 400 Pa [18–20]. Such conditions allowed for better yields and faster distillation of lactide.

Regarding the recrystallization yield during the purification process, limited information is available in the literature. For instance, Sanglard et al. [14] researched the recrystallization of lactide using different solvents, reaching a recrystallization yield using ethyl acetate of approximately 70% [14], slightly higher than the results obtained in this investigation.

4.2. Characterization of Lactide

The storage conditions of lactide influence its purity. Prolonged exposure time to oxygen and moisture can lead to the conversion of lactide back to lactic acid [21]. Therefore, in this study, all synthesized lactide was stored in a vacuum-sealed desiccator to minimize such degradation. On the other hand, although a second recrystallization may improve purity, it inevitably results in a loss of 20–30% of lactide [4].

The purity of lactide (90.0 ± 1.79%) is in agreement with that in other investigations. Sanglard et al. [14] reported lactide purities ranging from 84% to 96%, while Feng et al. [15] achieved an average purity of 90.73% with a relative standard deviation of 1.5%. However, the research by Hu et al. [10], on which our methodology was based, yielded lactide with a purity of 95%.

Remarkably, the lactide remained stable with a purity of 88.3% after four months of storage under the described conditions. This is noteworthy compared to studies where lactide stored for only 14 days experienced a 5% loss in purity [21].

Regarding the thermal properties of lactide, the melting and decomposition temperatures ($T_m = 97.2$ °C and $T_d = 227.8$ °C) are in agreement with those in other investigations.

For example, Peñaranda et al reported $T_m = 95.65$ °C and $T_d = 195.12$ °C [22], while Nyiavuevang et al. reached $T_m = 94$ °C and $T_d = 240$ °C [23].

5. Conclusions

This study demonstrates the feasibility of synthesizing lactide, a crucial intermediate in the production of poly(lactic acid) (PLA), using zinc oxide aqueous nanoparticles as efficient catalysts. The use of these catalysts resulted in lactide yields of 77–80%, indicating the effectiveness of the catalytic system.

The purification process using ethyl acetate showed a yield of approximately 57%, consistent with prior research studies. The average purity of the obtained lactide was 90.0 ± 1.79%, affirming the effectiveness of the purification method.

Overall, this study contributes to the understanding of lactide synthesis using zinc oxide aqueous nanoparticles as catalysts and provides valuable insights into the production of high-quality lactide for PLA manufacturing. Further research can focus on process optimization and scale-up studies to enhance the efficiency and yield of lactide production, further advancing the development of sustainable and environmentally friendly biopolymers as alternatives to conventional plastics.

Author Contributions: Conceptualization, S.D. and M.O.; methodology, S.D., M.O. and D.R.; validation, M.O. and D.R.; thermal analysis, A.D., M.O. and D.R.; HPLC analysis, F.P.F.; writing—original draft preparation, S.D., M.O. and D.R.; writing—review and editing, S.D. and A.D.; supervision, S.D.; project administration, S.D.; funding acquisition, S.D. All authors have read and agreed to the published version of the manuscript.

Funding: This research was funded by Faculta de Ciencias Químicas—UNA and Red and RED CYTED ENVABIO 100 (Ref: 121RT0108) (interaction, ZnO nanoparticles acquisition and publication cost).

Institutional Review Board Statement: Not applicable.

Informed Consent Statement: Not applicable.

Data Availability Statement: The data presented in this study is contained within the article.

Acknowledgments: The authors gratefully acknowledge the Red Cyted ENVABIO100 121RT0108 for the financial support. S.Duarte would like to thank the PRONII (CONACYT, PY).

Conflicts of Interest: The authors declare no conflict of interest.

References

1. Butbunchu, N.; Pathom-Aree, W. Actinobacteria as Promising Candidate for Polylactic Acid Type Bioplastic Degradation. *Front. Microbiol.* **2019**, *10*, 2834. [CrossRef]
2. Nofar, M.; Sacligil, D.; Carreau, P.; Kamal, M.; Heuzey, M. Poly (lactic acid) blends: Processing, properties and applications. *Int. J. Biol. Macromol.* **2019**, *125*, 307–360. [CrossRef]
3. NatureWorks. Ingeo. 2022. Available online: https://www.natureworksllc.com/ (accessed on 19 October 2022).
4. Nuyken, O.; Pask, S. Ring-opening polymerization-An introductory review. *Polymers* **2013**, *5*, 361–403. [CrossRef]
5. de Albuquerque, T.; Marques Júnior, J.; de Queiroz, L.; Ricardo, A.; Rocha, M. Polylactic acid production from biotechnological routes: A review. *Int. J. Biol. Macromol.* **2021**, *186*, 933–951. [CrossRef]
6. Mehta, R.; Kumar, V.; Bhunia, H.; Upadhyay, S. Synthesis of poly(lactic acid): A review. *J. Macromol. Sci.* **2005**, *45*, 325–349. [CrossRef]
7. Hu, Y.; Daoud, W.; Cheuk, K.; Lin, C. Newly developed techniques on polycondensation, ring-opening polymerization and polymer modification: Focus on poly(lactic acid). *Materials* **2016**, *9*, 133. [CrossRef]
8. Vink, E.; Rábago, K.; Glassner, D.; Gruber, P. Applications of life cycle assessment to NatureWorks™ polylactide (PLA) production. *Polym. Degrad. Stab.* **2003**, *80*, 403–419. [CrossRef]
9. Xu, X.; Liu, L. A study on highly concentrated lactic acid and the synthesis of lactide from its solution. *J. Chem. Res.* **2021**, *10*, 856–864. [CrossRef]
10. Hu, Y.; Daoud, W.A.; Fei, B.; Chen, L.; Kwan, T.H.; Lin, C.S. Efficient ZnO aqueous nanoparticle catalysed lactide synthesis for poly(lactic acid) fibre production from food waste. *J. Clean. Prod.* **2017**, *165*, 157–167. [CrossRef]
11. Tefara, S.F.; Jiru, E.B. Lactide synthesis via thermal catalytic depolymerization of poly lactic acid oligomer using ZnO nanoparticle dispersion. *J. Polym. Res.* **2023**, *30*, 146. [CrossRef]

12. Dong, K.; Kim, D.; Doo, S. Synthesis of lactide from oligomeric PLA: Effects of temperature, pressure, and catalyst. *Macromol. Res.* **2006**, *14*, 510–516.
13. Garlotta, D. A literature review of poly(lactic acid). *J. Polym. Environ.* **2001**, *9*, 63–84. [CrossRef]
14. Sanglard, P.; Adamo, V.; Bourgeois, J.; Chappuis, T.; Vanoli, E. Universities of applied sciences. *Chimia* **2012**, *6*, 951–954. [CrossRef]
15. Feng, L.; Gao, Z.; Bian, X.; Chen, Z.; Chen, X.; Chen, W. A quantitative HPLC method for determining lactide content using hydrolytic kinetics. *Polym. Test.* **2009**, *28*, 592–598. [CrossRef]
16. Cunha, B.; Bahú, J.; Xavier, L.; Crivellin, S.; de Souza, S.; Lodi, L.; Jardini, A.; Maciel Filho, R.; Schiavon, M.; Cárdenas Concha, V.; et al. Lactide: Production Routes, Properties, and Applications. *Bioengineering* **2022**, *9*, 164. [CrossRef]
17. Marchesan, A.; Motta, I.; Filho, R.; Maciel, M. Simulation of Multi-stage Lactic Acid Salting-out Extraction using Ethanol and Ammonium Sulfate. *Comput. Aided Chem. Eng.* **2020**, *48*, 1645–1650.
18. Zhang, Y.; Qi, Y.; Yin, Y.; Sun, P.; Li, A.; Zhang, Q.; Jiang, W. Efficient Synthesis of Lactide with Low Racemization Catalyzed by Sodium Bicarbonate and Zinc Lactate. *ACS Sustain. Chem. Eng.* **2020**, *8*, 2865–2873. [CrossRef]
19. Huang, W.; Qi, Y.; Cheng, N.; Zong, X.; Zhang, T.; Jiang, W.; Li, H.; Zhang, Q. Green synthesis of enantiomerically pure l-lactide and d-lactide using biogenic creatinine catalyst. *Polym. Degrad. Stab.* **2014**, *101*, 18–23. [CrossRef]
20. Upare, P.; Hwang, Y.; Chang, J.; Hwang, D. Synthesis of lactide from alkyl lactate via a prepolymer route. *Ind. Eng. Chem. Res.* **2012**, *51*, 4837–4842. [CrossRef]
21. Glotova, V.; Zamanova, M.; Yarkova, A.; Krutas, D.; Izhenbina, T.; Novikov, V. Influence of Storage Conditions on the Stability of Lactide. *Procedia Chem.* **2014**, *10*, 252–257. [CrossRef]
22. Peñaranda, J.; Agudelo, C.; Zuluaga, F.; Valencia, C. Synthesis of poly (lactic acid) and production of scaffolds by electrospinning. *MOJ Proteomics Bioinform.* **2016**, *4*, 334–338.
23. Nyiavuevang, B.; Sodkampang, S.; Dokmaikun, S.; Thumanu, K.; Boontawan, A.; Junpirom, S. Effect of temperature and time for the production of polylactic acid without initiator catalyst from lactide synthesized from ZnO powder catalyst. *J. Phys. Conf. Ser.* **2022**, *2175*, 012042. [CrossRef]

Disclaimer/Publisher's Note: The statements, opinions and data contained in all publications are solely those of the individual author(s) and contributor(s) and not of MDPI and/or the editor(s). MDPI and/or the editor(s) disclaim responsibility for any injury to people or property resulting from any ideas, methods, instructions or products referred to in the content.

MDPI
St. Alban-Anlage 66
4052 Basel
Switzerland
www.mdpi.com

Biology and Life Sciences Forum Editorial Office
E-mail: blsf@mdpi.com
www.mdpi.com/journal/blsf

Disclaimer/Publisher's Note: The statements, opinions and data contained in all publications are solely those of the individual author(s) and contributor(s) and not of MDPI and/or the editor(s). MDPI and/or the editor(s) disclaim responsibility for any injury to people or property resulting from any ideas, methods, instructions or products referred to in the content.

www.ingramcontent.com/pod-product-compliance
Lightning Source LLC
LaVergne TN
LVHW070042120526
838202LV00101B/389